THE LOST MESSAGE

How Crypto Drifted from Freedom to Speculation—and What Remains

Jamil Hasan

First Edition · 2026
Published by Crypto Hipster Publications

ISBN: 978-1-972991-01-5

Cover design by Jamil Hasan
Interior design by Jamil Hasan

Printed in the United States of America

For my wife, Amanda—who kept me grounded in
what matters,
when everything else was noise.

The Lost Message

Crypto was supposed to be about freedom. Not price. Not speculation. Not charts.

It was supposed to be about control. Who has it? Who doesn't? And what happens when that changes?

Somewhere along the way, that message got lost. Or at least it feels that way.

If you look at crypto today, it doesn't look like a revolution. It looks like a market. A fast-moving, attention-driven system where everything is measured, compared, and reacted to in real time.

Where value is defined by movement. Where participation is driven by opportunity. Where meaning gets squeezed into numbers.

But that's not where this started. This started as a response:

- To systems that restricted access.

- To structures that concentrated control.
- To a financial reality that most people accepted but very few understood.

Bitcoin wasn't created to outperform assets. It was created to replace the traditional banking system. And yet today it's often talked about as just another asset.

This book is not about explaining crypto. It's not about predicting markets. And it's not about telling you what to buy.

It's about something else.

It's about what this space was meant to be.

Through conversations with builders and people working beneath the surface of price and narrative, this book tries to reconstruct something that still exists but is harder to recognize:

- ➢ A system problem.

> ➤ A purpose problem.

> ➤ An access problem.

> ➤ A culture problem.

Four layers of the same reality.

If you're here for quick answers, this won't give them to you. But if you've ever felt like something about this space doesn't fully make sense... If you've ever wondered why the conversation feels different from what it was meant to be... If you've felt the gap between what is happening and what actually matters...

Then you're already closer to the message than most.

The question isn't whether the message is still there.

The question is whether you're willing to look for it.

Preface

I'm Jamil Hasan. People call me the Crypto Hipster. I gave myself that name. I own it. It started as a joke. A meme. Then it became a signal. Not rebellion for rebellion's sake. Rather, a way of seeing things.

Crypto moves fast. Sometimes, a topic I spent several podcast episodes fully understanding in the first week completely changed the next week.

There is a tremendous amount of noise in the market and not enough uncovering of the true signals. I aim to change that.

Discussion about the price of Bitcoin, meme-coin hype, 1000X speculation, and narrative repetition never truly interested me. I wanted to find out what genuinely matters instead.

So, I started the Crypto Hipster Podcast in February 2021. Not to chase headlines. Instead, to have actual conversations with people building this space.

Founders. Creators. Thinkers...people trying to reshape systems.

Crypto Hipster became more than a podcast. Nine seasons. Over 580 conversations. Hundreds of written works.

Then I hit an uncomfortable fact:

Curating discussions is not the same as creating ideas.

For years, I documented conversations. Published transcripts of all my podcasts. Built an archive. A colleague told me that if I wanted to be an influential leader, I would have to do more than simply host conversations. I would have to move the conversations forward.

That person was right.

This book, which is the first of sixty-seven Crypto Hipster Curtain Calls compilations spanning two-hundred eighty-one interviewees, is the first part of that shift.

Curtain Calls differ significantly from my first 399 books. It is not a collection of transcripts. It's an interpretation. A reflection. A synthesis of conversations, ideas, and lived experiences. It's where my voice meets the voices of the people I've learned from.

Crypto is not just technology. It's about power. About identity. About freedom. And above all, about people.

My path into this space came not from comfort. For the past five years, I have fought an aggressive desmoid tumor. What kept my mind busy and away from despair grew into something bigger. It became a way to process the world and my place in it.

This book blends two things:

1) The conversations I've had with builders of the digital economy.
2) And the lessons I have lived myself.

Conversations with others. And personal memoirs interwoven with them.

You'll see both here. You'll see ideas about decentralization and finance. About infrastructure and governance.

You'll also see stories. Resilience, failure, faith, relationships, recovery.

None of this technology exists in isolation. It reflects who we are. Too often, crypto gets reduced to price charts. The authentic story runs deeper:

a.) Whether these systems actually change anything or just recreate the same structures with a new label.

b.) Whether innovation leads to freedom or just a different method of control.

c.) Whether we, as individuals, choose to take part on purpose or just go with the current.

This book is my attempt to slow that down. To take hundreds of conversations and extract what matters most.

To connect patterns.

To ask better questions.

To share what I've learned. Not as answers. But as a framework for thinking.

This is not just my story. It's an invitation...an invitation to think differently. To question more. To build with intention.

I hope that if you see yourself in these pages, this book will serve as an inspiration for you to write your own story with more clarity, purpose, and conviction.

Table of Contents

Chapter 1—The Moment I Realized Something Was Off

I didn't come into crypto as a trader.

Not as an investor either.

Nor an enthusiast, advocate, or community builder.

I came in arrogance. And ignorance. I came in thinking I already knew it all.

I was wrong. I knew nothing.

For twenty years, I built databases. Structured systems. Organized information. I designed ways for data to move, to be stored, and to be accessed quickly.

That was my frame. I saw the world as inputs, outputs, rules, and constraints. Everything could be modeled and put into an architecture.

So, when I first found Bitcoin, I reduced it to something familiar: a public database.

That was my take.

Bitcoin was simply a distributed ledger. An immutable record. A way to store and verify transactions without a central boss.

Technically, it made sense. Elegant, even. But it felt contained. Understandable. Something I could map onto what I already knew. I thought I was looking at an innovation in data. I said to myself, this is just a database, and I know databases.

I did not know I was walking into something much bigger. It was the biggest miscalculation of my life.

A public database doesn't start a revolution. Bitcoin wasn't just a database. It was a signal. And I read the wrong layer. I didn't realize I was not just seeing new technology. Instead, I was stepping into a cultural war. A financial revolution. A generational split. None of that was obvious from where I stood.

At AIG, Bitcoin wasn't a serious topic. It existed, sure, but it didn't penetrate. It didn't force engagement. It was kept at the edge. Acknowledge it and move on. Dismiss it. Defer it. No consequence.

In hindsight, that absence mattered. What isn't discussed isn't understood. And what doesn't get understood is easy to underestimate.

I found Bitcoin right after I got laid off in June 2017. That timing changed things. When you're inside a system, you accept its edges. When you're out of it, you question them.

Curiosity led me in. Something to explore. Something to understand. Then it turned into something else.

The more I looked, the less it matched my first idea. It wasn't just a database. It challenged how value databases had always been controlled. It wasn't just a ledger. It removed the need for trust in the institutions that ran those ledgers. It wasn't just a technical trick. It was a philosophical shift. Once I saw that, I couldn't go back to my old interpretation.

It wasn't the price that drew me in. Not the idea of a new asset class or a new way to make returns. It was the sense that this space was asking a different question. Not how to make old systems better. But whether those systems needed to exist as they did.

I didn't have the words then. But I felt it. Conversations in crypto differed from anything I'd seen. Not just product building or company scaling. People were rethinking things most never questioned. Money. Ownership. Trust. Coordination. These weren't features. They were foundations. And for the first time, those foundations felt open to redesign. That hooked me.

Something less philosophical hooked me, too. My initiation fee.

Shortly after I stepped into what felt like the wildest version of the Wild West in 2017, I met a man online who offered to help me understand Bitcoin. I said yes. We talked for about a week. Long enough for me to feel like I was learning something real. Long enough to trust him. Then I got on a plane to Las Vegas. I met Morgan on Fremont Street in downtown Las Vegas. He showed me the Bitcoin ATM he had installed at a casino. Walked me through how it worked and how people use it. How value moved in and out of something that, until then, had only existed on my screen. Later that night, we went to a steak dinner. He paid for it in Bitcoin. Simple moment. But it changed something.

Until then, Bitcoin had been conceptual. An idea. A system. A structure I was trying to make sense of. That dinner made it real. Not just something to study. Something you could use. When I got back home, I did what many people did in those early days. I leaned in. I called myself a Bitcoin advisor. Looking back, I didn't know enough to justify that title. But in 2017, almost no one did. And the space moved fast.

My LinkedIn messages started coming in. My phone kept ringing. People were asking questions. Trying to figure it out. Trying to get ahead of something they felt mattered, even if they didn't quite know why. I landed no business from my Bitcoin advisory practice. But I found myself in situations I wasn't ready for.

Once I got pulled into a dispute between a Bitcoin exchange in India and its community. People were saying they'd been ripped off. Frustration was rising. Trust was breaking down in real time. I wasn't fixing anything. I was just watching, trying to find the words for what I was seeing.

This space made not only opportunity. It made conflict. Human conflict. Trust. Mistrust. Expectation. Breakdown.

Two months later, my father died. I was in shock. Everything stopped. The rest of the noise fell away. One of the last things he told me before he died was that I should build something that was me. Not work inside something. Not to attach myself to something. Build something that reflected who I was. I didn't get it then. I did nothing about it. But it stuck with me.

In the middle of crypto and conversations and noise and opportunity, that line kept coming back. Not as instruction. More like something unresolved.

Later that year, I sat on a panel about a comment Jamie Dimon made in September 2017, calling

Bitcoin a fraud. It felt like another conversation. Another moment in a fast-moving space. I don't remember exactly what I said. I remember the room moving on.

A few years later, the host of that panel contacted me. She said something I had said on the panel had reached a group of women in Pakistan. That those women had used those words to stand up against the Taliban. And forced them to leave their community. I didn't see it happen. I didn't witness it. I didn't even know until years later. But it happened. That changed something again. It made clear what I hadn't fully seen before:

You don't always see the impact of what you say.

That doesn't mean it's not happening. And in crypto — where ideas move faster than systems and narratives spread faster than understanding — that impact multiplies in ways you can't easily track or predict.

That was the environment into which I was stepping. Not just new technology. Not just a new market.

A system where ideas, money, belief and behavior all interacted at once. Consequences that weren't always visible.

At the time I didn't connect all the moments. The Las Vegas dinner. The messages. The confusion. The conflict. My father's words. The panel. The surprise impact years later. They felt separate. But looking back, they weren't. They were signals. Different expressions of the same shift. A system being built. A narrative forming. A space expanding faster than could be understood. And somewhere inside it all, something important was already beginning to drift. I just didn't have the language for it yet.

Shortly after my father died, I paid another initiation fee. This one wasn't philosophical. It was financial. And it burned. I was in shock for months. Not the kind you show. The kind that makes everything feel distant. Decisions felt muted. Judgment dulled. I wasn't thinking clearly.

Someone reached out on Facebook. He said he ran a legitimate crypto exchange. Spoke as if he knew what he was doing. Made it sound simple. If I wanted to make money, he said, I should invest with him.

At another time, I might have slowed down. Asked the right questions. Been skeptical. I didn't. I sent $10,000.

At first, it looked like I'd made the right call. I watched the account grow. Ten thousand became twenty. Twenty became thirty. Thirty became fifty. That growth felt like proof. Not only that, the investment worked, and I was getting it. I was learning something.

After a few months, I tried to withdraw. And then everything changed. I tried to access the funds. Nothing happened. I followed up. Asked questions. I tried to understand the process. Then I was told I couldn't withdraw because I had referred no one else to the platform. That was the moment. Not when the money disappeared. But when the logic broke.

Systems aren't supposed to work like that. The more I questioned, the more it unraveled. Until there was nothing left to question. I was blocked. The account vanished. And the money was gone. Just like that. No escalation path. No customer support. No authority to appeal to.

Nothing.

That was my second initiation fee. It taught me something I didn't fully get until later:

Not your keys, not your crypto.

It felt like a loss. Looking back, it felt different. More like a confrontation. Between ownership and access. Between seeing something and controlling it. Systems with no accountability don't fail loudly. They disappear quietly.

That experience changed how I handled things after. Not right away. But for good. It forced a level of understanding that no conversation could give me. It pulled decentralization out of theory and put it into consequence. Self-custody stopped being an idea. It became a requirement. It also showed me something else. This space wasn't just about innovation. It was about responsibility. And that responsibility wasn't optional. It was part of the system.

When I look back, that moment sits with everything else. Dinner in Las Vegas. The early conversations. The confusion. The unexpected impact of words. They weren't separate. They stacked up. Each one showed another side of the system. The promise.

The risk. The potential. The consequence. And in the middle of all that, something else was forming. Not clear. Not finished. But clear enough to see this space was more than it first seemed. And that was it seemed to be was already drifting.

I didn't arrive in 2017 with a technical plan. By 2021, I had more questions than answers. Those questions turned into conversations. Hundreds of them. Founders. Entrepreneurs. Independent builders. People working on stuff that was early, hard to explain, often misunderstood, but tied to something deeper than a surface opportunity.

Those talks became the Crypto Hipster Podcast. They also became my lens. Talk to enough people building at the edge and patterns show up. Not right away. Not neatly. But gradually. You hear the same ideas from different voices. The same worries are framed differently. The same drives are present even when the projects look nothing alike.

What stood out early was how consistent the patterns were. People weren't just building to build. They wanted to fix things. They talked about control. Systems that can cut you off without warning. They talked about money that can be inflated, manipulated, or devalued and you can't do anything about it. They talked about coordination. Resources that exist but never reach where needed. Inefficiencies from misalignment, not scarcity.

They talked about access. Open systems that feel closed if you don't know how they work. Participation that requires belief, not just availability. They talked about community. What happens when people gather without logical structure. How things drift when no one is holding them in place. None of this was about price. That's why what happened next felt so wrong.

While people were building and solving genuine problems, the story around crypto changed. Slow at first. Then fast. It turned into a price story. Not completely. Not always stated. But it became the reference point. Price became the entry point. Everything else got measured against the price of Bitcoin.

Projects were judged by token moves, not what they enabled. Conversations that started with structure ended in speculation. Ideas meant to reshape systems got filtered through what they might return. At first, that seemed okay. Validation. More attention. More participation. More capital.

From the outside, it looked like progress. But something felt off. Not one moment showed it. A pattern did. A growing gap between what people were building and what people cared about. I'd talk to a founder working on something structural that could change whole systems. Then I'd watch how that project was discussed publicly. Not about what it enabled. About what it might do to price. The translation was small. But constant. A coordination system turned into an opportunity. A tool for access became an asset. A transformative idea turned into a trade. Same words. Different meaning.

I thought it was a phase. Something to correct as the space matured. But the opposite happened. The bigger the space got, the more that lens dominated. And the louder it got, the harder it was to hear anything else. That's when the question formed. Not fully clear. But persistent.

What happened to the original message?

When I entered the industry, I was inspired by the three blockchain promises:

Banking for the unbanked population.
Identity for unidentifiable people.
A Voice for the Voiceless.

I saw myself addressing that promise of voice.

I didn't feel that I had been disproven. People hadn't stopped building. The problems were not solved. Something else had taken the place of that message. Though the message hadn't vanished, it had been displaced. Not removed. Not rejected. Just pushed to the background. Replaced by what's easier to measure. Easier to react to. Easier to talk about in real time: price.

Once that happened, everything shifted with it. Conversations changed. Not so much the words as the way they were read. Participation changed. Not how many, but why. Even how I listened changed. That surprised me. I wasn't only watching the shift. I was in it. Filtering things the same way. Looking at the same signals. Interpreting ideas through the same lens. Once I saw I couldn't unsee it. The problem wasn't that the message was gone. It was harder to hear. Not because it changed. Because the surrounding environment did.

The signal was still there. But the noise got louder. That realization grew. Across talks. Across patterns. Across moments when something felt slightly off and I couldn't explain why. Eventually, it became clear.

The space hadn't lost its purpose. It had shifted its attention.

Once attention moves, everything follows. That's what this book is about. Not a primer on crypto. Not a prediction of what comes next. But digging back to something that still lives below the surface.

The conversations that follow aren't new. They aren't reactions to the latest headlines. They aren't shaped by cycles or narratives. They reflect what this space was always meant to tackle.
To see that, you must listen differently. Not for what is moving. But for what is being built. Not for what's happening. But what it means. If you listen

the same way everyone else does, you'll hear the same thing everyone else hears.

Shift how you listen and something else appears: a pattern. And that pattern points to the core idea behind what follows.

The message didn't disappear.

We just stopped listening to it the way it was meant to be heard.

Chapter 2—What These Conversations Reveal

The conversations did not change. That made the shift hard to notice at first. The people I talked to— founders, builders, creators—were still solving the same problems. Systems, access, and coordination. How to build something that doesn't just tweak what exists but rearranges it.

Their language remained the same. Their intent too. Their work stayed tied to real-world effects and to the ideas that drew people here in the first place.

And yet the way those conversations landed had shifted. Not inside the room. Outside of it.

Two versions of the same space existed at once. One you could see. One you couldn't. The visible one moved fast. Immediate. Driven by price, narrative, and speculation, and always trying to explain what was happening right now. It rewarded the reaction.

It rewarded attention. It made a lot of things feel urgent, even when that urgency was temporary. Time felt compressed. Everything was judged by what it did at that moment.

The other one was slower. Under the surface. Not hidden, just not trying to be loud. It showed up in long conversations, in early systems, in messy problems without deadlines. It needed patience and context, and a willingness to sit with ideas that aren't made for fast takes. No constant feedback. No quick rewards. You had to understand before you reacted.

Both were real. But one was louder. Over time, the gap became more than speed. It became about how things were read. When the primary frame people rely upon to understand a system changes, the meaning of everything inside that system shifts too.

Talk about infrastructure becomes talk about upside. Talk about access becomes talk about opportunity. Coordination becomes an asset to allocate. Structural ideas are read as financial. Transformation gets reframed as a potential return.

Same words. Different meaning.

This is not people failing to explain themselves. It's framing doing work. People read systems through the signals they see most. If price and volatility and short-term moves dominate, everything else gets squinted at through that lens. Deeper ideas are not ignored. They get translated into something more familiar. People fit what they hear into what they already know. And when that translation happens again and again, it becomes the default way to read things.

A system stops being judged by what it enables. It is judged by what it returns.

What I noticed repeatedly was not contradiction but convergence. Different people. Different backgrounds. Different domains. The same basic ideas showed up repeatedly, often with no one coordinating or citing each other. Projects were not building the same frameworks and not agreeing on a single story, but were describing the same patterns and constraints from different angles.

Regarding control, access, purpose, and responsibility, each conversation came at those themes differently. But they lined up. Not as separate problems, but as different sides of the same thing.

Dr. Karl Huber looks at the system's base: money itself. And money is not neutral; rather, it is a tool for control. He focuses on how value is made, shared, and reached. He sees the link between governments and banks as a central break. A closed loop that concentrates power, causes inflation, and makes participation conditional instead of sovereign. For him, crypto isn't a market category. It's a response. It is a way to drop dependencies and to make ownership direct and access hard to cut off. His work pushes against the market's story. It pushes against the structures in which markets live.

Violet Abtahi looks at purpose, not structure. She cares about what systems produce for people. She questions whether stability alone is enough. A system that lets people function might not let them grow. She moves the problem from scarcity to coordination. It's not just about lacking resources. It's about resources moving wrongly. To her, crypto's promise isn't only decentralization. It's waking up unused value, building systems that lift people past survival, and matching resources with opportunity in ways old systems don't. But that promise isn't automatic. It needs intention. Builders must know what they aim to enable.

Claire Ross Brown brings us to entry, where ideas hit reality. She points out something often missed: accessibility is partly psychological. The system can be open in principle and still feel closed. People don't join just because the gate is open. They join because they believe they belong. The barrier is not always complexity. It's doubt. Not feeling qualified. Assuming others already get it. Without that psychological permission, access remains partial. The system is there, but not fully embraced.

Anjali Young fills out what happens after people enter. She looks at behavior, community, and culture. Her view comes from long experience with online communities that existed before crypto. She reminds us that open systems don't automatically steer toward excellent outcomes. They mirror the people inside them. Without boundaries, leadership, or incentives that align, communities drift. They don't crash overnight. They fray. Over time, that fraying shapes what the system becomes. Decentralization here is not dumping responsibility. It's spreading it.

Together these views make a fuller picture than anyone alone. A system problem. A purpose problem. An access problem. A culture problem. These are four ways of describing the same reality...a reality that is hard to see live because of the environment this all happens within.

While these talks occur, the loudest signals point somewhere else. Attention goes to what moves, what updates, what you can measure right now. Things that don't give quick feedback fade. Not because they matter less, but because they don't respond to attention in the same way. That makes a gap, not between what exists and what could be, but between what is being built and what is noticed.

That gap is where the message gets muffled. Not gone. Just not the loudest voice. And that displacement doesn't feel like losing something. It looks like progress. The system grows. More people join. Capital moves. New projects pop up. Activity speeds up.

From the outside, it seems like forward motion. Movement feels like advancement. But motion alone doesn't tell you the direction. A system can move and still drift from its purpose. It can grow in ways that make sense inside the system but not outside of it. It can optimize the signals it receives without asking if those signals match what really matters. And the system keeps working. It just makes something else.

Interpretation matters. How people read these conversations shapes what gets reinforced. Prices feed the dominant story, which gets filtered down to what they mean for movement and returns. Structural meaning shrinks into financial meaning. Meanwhile, the systems show something different. No predictions. No neat conclusions. But structure.

What follows is not a tidy, single argument. It's an attempt to rebuild that structure. To align these views with one another. To make the pattern visible when you see them together, not alone. To move from reaction to a bit more sense. To give a frame that lets the quieter signal be heard again.

Once that frame exists, one question sits under everything: What was this space meant to become?

This displacement sticks because it grows from inside, not because someone outside pushed it. People respond to what's visible. Price, momentum, narrative.

In a world of too much information and too little attention, the easiest signals rise.

Price does that. It shrinks complexity to a single number. It points without context. It becomes the anchor for interpretation.

Anchors have effects. Once reading is fixed to one dimension, everything lines up with it. Information gets filtered. Ideas get judged by how they move price. Even systems meant for other things get recast in terms of price influence. Over time, the view narrows. Not because other options vanish, but because they stop being reinforced by the loud signals. That's how narratives form. Not by everyone agreeing aloud, but by attention aligning repeatedly. When enough people read the system the same way, that reading sustains itself. No one needs to force it. Repetition, visibility, and feedback do the job.

Once a narrative gains that traction, it's hard to move, even if it leaves things out. Changing a narrative takes more than additional facts. It takes shifting attention. Attention acts like selection. It chooses what gets amplified and what stays on the side. It shapes not just what people see but what they think is worth looking at. This creates a hierarchy of what matters that may not match the system's actual shape.

In crypto, that hierarchy favors immediacy. Things that update fast outrank what develops slowly.

Easy-to-measure things beat things that need thought. Things that spark reaction win over things that require understanding.

This pattern is bigger than just crypto. But in crypto, it matters because the system's most critical parts, which are structure, purpose, access, and culture, don't send quick signals. They need a long attention span. That mismatch makes what matters and what's noticed different. The system is judged by signals that miss its principal features.

When evaluation is wrong, development follows the wrong map. Builders go after attention. People play where it looks like things matter. Money flows to the visible. In the long-term, the system optimizes for what's easiest to amplify. That's how displacement gets built in. Not one big switch. Lots of small moves stacking up. And the message still doesn't feel missing altogether. Because pieces of it remain.

Those pieces remain in conversations that pop up briefly and then get folded back into the loud story. They remain in projects trying to work on structure but get judged by market moves. Pieces can be found in ideas people nod at but don't prioritize.

The signal is not gone. It's diluted. And what is Diluted has not vanished. The meaning stays, but clarity drops. And when clarity falls, recognition becomes harder. So, interpretation matters again.

While attention decides what is seen, interpretation decides what is understood. Interpretation can change. You don't have to throw out the visible layer. Put it in context. Remember, price is a signal, not the complete system. Movement is an indicator, not the goal. Activity is a metric, not the purpose. Make that split and the structure under the noise gets clearer.

The conversations ahead are not random takes. They are parts of that structure. Each one captures a slice of the same system. Each one helps make sense of what gets reduced to surface signals. Read together, they don't fight. They line up. And when they line up, the scattered bits form again. Not a new idea. A clearer view of one that was already there.

If the message feels difficult to find, it's not because it disappeared. It's because we stopped listening for it.

Chapter 3—Financial Freedom Requires a New System

Dr. Karl Huber is the co-founder of the Quai Network, a new Layer 1 network of blockchains that scales to handle all human commerce while maintaining true decentralization. Dr. K's vision for a trustless decentralized replacement for money began in 2018 with a National Science Foundation Research Grant. Dr. K has a PhD in engineering, formerly worked at ConsenSys, and has been active in crypto since 2012. He has authored or co-authored dozens of peer-reviewed articles and papers and has contributed to many groundbreaking developments in the crypto space.

Podcast Guest: Dr. Karl Huber

Podcast Date: March 17, 2025, Season 8

Podcast Link:

https://open.spotify.com/episode/0vFk6dvOhLnQ BZzhFaa4Z4?si=m4r7ejdvTUCkrSgwUpak6w

Before crypto became a market, it started as a response. Not to the price of Bitcoin. Not to opportunity. And not to speculation. It came from a structural problem most people live inside without naming.

The modern financial system feels given. Like gravity or time. People learn how to take part in it. How to earn money. How to spend. How to save. Rarely how to question it. On the surface, it runs smoothly. That smoothness breeds passive trust. But stability in a system where the rules can change is not neutrality. It is managed. And what is managed is controlled.

For Dr. Karl Huber (Dr. K), the entry into crypto was not tech fascination or a desire to make money. It was a conceptual fracture. He saw money, something most people treat objectively, as a system of choices. What hooked him was the idea of "separating money from the state... creating a voluntary economic system." Once you see that, it is hard to unsee.

If money is tied to the state, then participating in the economy is not purely voluntary. It is mediated by structures outside your control. And if those structures can change the rules, including how money is made, who gets it, and how it flows, then the system becomes a type of power, not just a medium of exchange.

Most people never feel that power directly. They feel outcomes. Prices go up. Purchasing power falls. Savings erode over time, and it is hard to pin that on one cause. Those shifts are often called natural cycles. But the mechanism is more direct. As Dr. K says, "They can arbitrarily print... and inflate our purchasing power away." This is not an oddity. It is a feature of a system where money supply expands through policy choices that hit people unevenly.

The gap between creating money and feeling the effects hides the link. The result is steady. Value often moves away from individuals and toward the structures that control distribution. At its center is a relationship people rarely describe plainly. Governments and banks align. "This monopolistic relationship between the banks and the government removes the power from the people," Dr. K explained.

Being monopolistic does not mean one actor. It means a closed loop. Governments issue currency and set policy. Banks distribute capital and gate access. Together they make creation, distribution, and control tightly coupled. You can take part, but you are not sovereign. You operate inside a system you do not define.

Over time, that difference matters. Usually, this structure remains abstract. It does not need to shout. It sits in the background of daily life. But sometimes the system reveals itself.

Dr. K points to the Canadian trucker protests as one of those moments, not for politics, but for what it showed about financial infrastructure. "They froze the bank accounts of anyone who donated... That tells you why you need money independent of the state." What mattered was the mechanism. Access could be restricted without physical force and through control of the system. Money was not destroyed; it was made inaccessible. Ownership did not vanish, but was overridden. That changes how ownership looks.

Most people think ownership is absolute. You have something, so you control it. But in a system where access can be revoked, ownership becomes conditional. It exists while the system permits it. Conditional ownership differs significantly from real ownership. It depends on things outside the individual. The system does not need to use that power often for it to matter. The possibility alone reshapes the relationship.

Crypto grew out of this tension. Digital assets were not meant as a speculative thing, but to remove dependency. The aim was not a new asset. It was a new form of money, which could not be censored, arbitrarily inflated, or seized by a central authority. "The important piece is making a digital currency which reinforces the liberties of individuals," Dr. K says.

That framing is precise. The point is liberty. To create a system where the individual is the ultimate point of control. Crypto is less about flashy tech and more about architecture. It is about redesigning how value is created, stored, and moved.

Bitcoin captured that idea at the start. Designed as peer-to-peer electronic cash and direct, permissionless transactions between people, Bitcoin's meaning was not digital scarcity alone. It was removing intermediaries from the cash exchange altogether. For the first time, value could move without central approval. That was the breakthrough. Not price. Not mass adoption.

Functionality.

Over time, the role of Bitcoin and crypto shifted. Not the code, but how people used and understood it. "Bitcoin has failed as peer-to-peer electronic cash," Dr. K observes. Instead of becoming common money, it turned into digital gold. A store of value to hold, not a system to use. Some call that progress. But it is a trade-off. A system built for utility becomes an asset. Its major job changes. It moves from enabling transactions to generating returns.

And when that happens, incentives change. Speculation follows. Price becomes the main signal. Not because it tells you the most, but because it is the loudest. It updates constantly. Easy to track. Creates urgency. Attention is scarce. Visibility wins. And when focus goes to price, other things fade, such as, most importantly, purpose, design, and long-term consequences. Not because they matter less. Because they are less immediate.

This is where the drift starts. When crypto is talked about mainly as a price, it is no longer judged as a system. It is judged as an asset class. That brings in a different participant: people who care about allocation, returns, and timing. Not about building, usability, or infrastructure. As participation changes, development follows. What gets built reflects what gets rewarded.

Big institutional interest speeds this up. As crypto gains value, it draws funds, banks, and even governments. "Bitcoin becomes captured by BlackRock, Fidelity, and potentially the government itself," Dr. K notes. Not a hostile takeover. Just capital seeking exposure. But a new dynamic appears.

The more crypto folds into the old financial system, the more it looks like it. Custodial services, regulations, and financial products make access easier. They also bring intermediaries back. And with intermediaries comes control. That tension sits at the heart of crypto's path. Integration can mean more adoption and legitimacy. It can also erode the independence that defined it.

If crypto becomes fully part of traditional finance, it may win as an asset class and lose as an alternative. Short term, you might not see the difference. Long-term, it matters. Dr. K warns about an "Orwellian panopticon." Not because someone plans it, but because of structure. If money goes fully digital and those digital systems are centralized, then control reaches into every transaction. And if control over transactions is centralized, participation becomes conditional. The system does not need to block everything. It only needs the power to do so.

Still, the original problem is not solved. The forces that pushed people toward crypto, like centralized control, inflationary dynamics, and conditional access, are still here. They shape outcomes in ways that are often invisible but consequential. The need for an alternative hasn't shrunk. If anything, it looks clearer. What changed is the focus.

Attention moved from structural problems crypto aimed to fix toward surface-level things like price and speculation. The conversation shifted. Not because the original ideas lost force. Because something else was easier to measure and react to. That shift is hard to notice because it does not feel like a loss. It feels like participation. People join because the price is visible. It is the clearest sign of activity in a messy scene. Once you are in, that signal reinforces itself. Gains confirm attention while losses demand more of it.

The system trains people to watch and react through that lens. Over time, the lens becomes opaque. But what you gain in speed, you lose in depth. Price alone does not explain much. It shows movement, not meaning. It says something is happening without saying what or why. Yet when price dominates, it crowds out deeper questions.

Structure, control, and long-term viability get pushed to the margins. Not because they are unimportant, but because they take time, attention, and patience. And patience is what the system wears away. In a market driven by immediate signals, time compresses. Decisions get shorter. Feedback loops tighten. The gap between action and reaction shrinks. In that squeeze, the skill of thinking in systems becomes harder, nearly impossible, and less common.

The environment rewards speed, not reflection. Reaction, not building. This is not only about crypto. It is how attention works these days. But in crypto, it matters more because the space depends on long-term structural change. It needs people willing to build things that won't pay off right away.

Tools need understanding before they are useful. And they require a preference for independence over convenience. These are not short-term incentives. When short-term incentives rule, long-term systems struggle. That creates an inversion. The very forces crypto aimed to escape, such as extraction, short-term optimization, and concentration of attention, reappear within it. This reappearance is not because someone planned that, but because incentives push in that direction.

What is easy to measure becomes what you prioritize. What you prioritize becomes the focus of building. There is also comfort in that shift. Structural issues are slow, complex, and uncomfortable. They force you to admit that systems you rely on may not serve you. They show control in places you cannot change. Genuine change takes hard work and may bring no quick wins. Price is simpler. It moves. It gives direction, even if it is volatile. It gives people something to hold. And it lets them avoid the deeper questions that started the space originally.

Dr. K spoke about how those questions do not vanish. They sit there, unresolved, under the surface. They keep shaping outcomes even when people stop talking about them. Each cycle, every change in attention, and every wave of new participants runs into the same reality. The system that crypto addressed still exists, and the basics remain unchanged.

And because the basics are still the same, the message cannot be lost forever. It is tied to a problem that remains unsolved, though hidden, put aside, and misread. But as long as the conditions that created it stay in place, the need for it stays too. And that need comes back. Slowly or suddenly, it comes whenever the limits of the current system become too clear to ignore.

The question is not whether the message exists. It is whether enough people will go back to it. Not on price level. But at the level of infrastructure. Not as market participants. But as system builders. Because that gap between roles is where things change.

The system never stopped being the problem. We just stopped paying attention to it.

Chapter 4—Decentralization Only Matters If It Creates Abundance

Violet Abtahi is the founder and CEO of Platonic. Violet is a serial entrepreneur and investor with profound expertise in finance and blockchain technology. She has over two decades of experience in the financial sector, blockchain, and crypto ecosystems. Abtahi is a visionary leader on a mission to bridge the public and private blockchain worlds, shaping the next generation of financial infrastructure.

As the founder and CEO of Platonic, Violet spearheads the development of enterprise-grade L1 blockchain solutions, addressing real-world challenges in financial services. Through her leadership, Platonic is dedicated to building the next generation of financial markets infrastructure in the issuance and distribution of real-world and native assets on chain, focusing on scalability, security, and innovation.

Besides her role at Platonic, Violet is also the co-founder and president of Enya Labs, a leading blockchain developer, and the co-founder of Boba Network, a prominent public blockchain layer 2 solution. Her diverse portfolio of ventures reflects her commitment to driving innovation and advancing the capabilities of blockchain technology. Abtahi is not only a successful entrepreneur but also a dedicated investor and advisor to thriving projects with a profound purpose to serve the future of finance and humanity. Through her strategic guidance and support, she continues to shape the landscape of finance, fostering growth and innovation in the industry.

Podcast Guest: Violet Abtahi
Podcast Date: May 30, 2025, Season 8
Podcast Link:
https://open.spotify.com/episode/4vmThgt1RTp3Q4GzOoVinT?si=9ohd2KziT-6SF4k_Nztetw

Technology without purpose is just noise.

If the problem starts with how money is structured, it doesn't end there. Fixing the system doesn't automatically tell you what the system should do. A more efficient mechanism won't guarantee anything meaningful. Redistributing control doesn't suddenly make life better. It just makes other things possible.

What fills that space is not decided by code or infrastructure or decentralization alone. It's decided by intent. And it is decided by why people build, why they join, and why they try to steer things. When intent is vague or just assumed, systems keep running, but they drift. They grow without direction. They optimize without a clear aim. After a while, they look impressive on paper and feel empty in purpose.

Violet Abtahi looks at crypto from that layer. Not from how things work, but from why they exist. She starts with how people live inside systems that most take for granted. Most people don't choose the shape of their lives. They inherit it. They wake up. They work. They earn. They spend. And they do it repeatedly with little variation. It feels like progress, but it does not change the overall path.

Because the loop is stable. It gives continuity and predictability. It offers a kind of security. So it is rarely questioned in a way that changes the structure. It becomes normal and then internalized as the baseline of life. "Is life enough if we just wake up, work, make money, and repeat the loop again?"

Violet's question hits home because it's simple. It doesn't attack the system directly. It points to the assumption underlying it. Is repetition enough? Is mere survival, even a stable kind, the same as fulfillment? Would people pick that system if they knew there was an alternative?

Answering honestly is hard. The system stays not because it's best but because it's familiar. It sets the expected limits. Inside those limits, people learn to be efficient, even if that efficiency means less meaning. Violet calls this survival mode. The phrase fits because it captures the economic and cognitive sides. Survival mode isn't just about not having enough. It's about how scarcity shapes thought.

When attention goes to keeping things stable, earning enough, avoiding loss, and managing risk, there's less room to think beyond that. The mind adapts. Focus shifts to immediate outcomes. Tolerance for risk shrinks. Exploring things without a clear short-term return drops.

Violet says, "We need to move beyond survival mode... When we're in survival mode... our higher brain functions are offline." That is not a metaphor for being uncomfortable. It's describing how thinking narrows. When that happens across many people, the systems they live in reflect it. Innovation becomes small and iterative. Participation becomes reactive instead of creative. People optimize by keeping their place instead of questioning the place.

Structure does not have to crush creativity. It just needs to make creativity secondary to staying afloat. Most people call this scarcity. Not enough money, not enough opportunity, and not enough access all equate to scarcity. Hence, people stay constrained.

Violet flips the frame and pushes the conversation in a different direction. "I don't think we have a resource problem... we have a coordination problem." If scarcity is the issue, the fix is accumulation. More capital. More production. More expansion. But if coordination is the problem, then it's not about what exists. It's about how things are arranged. It's about the paths value takes, who gets access, and the friction that keeps resources from where they're needed.

The world is not empty. It's badly connected. Assets sit idle while demand exists elsewhere. Capital piles up in one place while another stays under-funded. People who could help are cut off from where help matters.

These are not supply failures. They are alignment failures. Old systems don't solve that well because they value stability over adaptability. They use intermediaries to manage distribution and fixed structures to keep order. Processes aim to lower risk rather than boost responsiveness. That makes them reliable, but also rigid because they don't re-route capital easily when conditions change. They add new players with friction. They don't develop fast when needs shift.

Over time, rigidity builds up. Inefficiencies normalize because they persist. Blockchain offers a distinct possibility. Not because decentralization is a slogan, but because code is programmable. Rules for distribution can live in code instead of being enforced by institutions. Value can move on conditions that are written, audited, and executed without one central boss. That makes it possible to build systems that are dynamic, not fixed, that react to inputs instead of sticking to set structures.

But having that possibility doesn't mean it will improve outcomes. Systems don't choose their own purpose. They do what the intent inside them tells them to do. And here's the hard part: intent is rarely spelled out. In a space that moves fast, where people race to build and scale and grab attention, purpose gets pushed aside. Speed and growth take over. The assumption becomes that innovation alone will fix things. But innovation is neutral. It widens what you can do. It does not pick which things you should do.

Violet assets that "if we are building technologies without thinking about that why, then we have lost the perspective." Perspective is not some vague ideal. It's a constraint. It guides decisions. It decides what to build, what trade-offs are acceptable, and what counts as a meaningful result. Without perspective, systems sprawl. They chase metrics that are easy to measure but not tied to long-term value. Growth becomes the target instead of the side effect of something deeper.

Instead, Violet recasts innovation around human experience. "Innovation... is about breaking through space-time barriers... so that we have more time to connect." That cuts through the abstraction. If innovation speeds things up, lowers friction, and widens reach, then ask what people get from that. Does it free up time for genuine connection or just cram more into less time? Does it make people closer or create shallow contact? Does it spread opportunities or concentrate them in new ways?

Faster systems do not automatically make things better. More efficient systems do not automatically make things fairer. Scale does not equal alignment. If we don't measure success by human outcomes—

time, connection, opportunity—then success gets reduced to surface metrics. Under all this is a deeper, nontechnical problem. "The core of our illness is disconnection." Disconnected from each other. From purpose. From systems meant to support rather than limit.

This disconnection doesn't always look like a breakdown. It can hide in systems that seem to work well. But it shows up in inefficiency, in inequality, and in that hollow feeling even when basic needs are met. And when tech is built on top of disconnection, it amplifies the problem. It speeds communication without deepening understanding. It increases access without creating alignment. It grows networks without building meaningful ties.

This is the drift from the earlier chapter, but in a new form. Dr. K pointed to the structural issue: the concentration of control in finance. Violet shows what happens when a fix is built without an obvious purpose. Instead of using crypto to improve coordination, unleash potential, and expand capacity, the focus shifts to price, speculation, and short-term gain.

Same tools. Different intent. And when intent changes, so do outcomes.

Price becomes the main signal because it's immediate and visible. It gives a sense of direction, even if that direction wobbles a lot. It makes feedback loops that reward attention and shape behavior. Over time, it replaces deeper evaluation. Systems get judged by growth instead of contribution, by activity instead of alignment, and by market moves instead of the problems they were supposed to solve.

That doesn't make the system fail right away. It keeps growing. Adoption goes up. Value ticks higher. But something quieter erodes. Purpose blurs. The link between what's built and why it matters gets harder to follow. The system becomes busy, not coherent. It produces results, but those results don't always connect back to the original aim.

Still, nothing essential is gone. The ability to coordinate resources in real time, to unlock untapped value, and to move people out of survival mode are still there. The tools exist. The infrastructure keeps getting better. The potential remains.

But the question is whether we're pointing the system at those ends. Reorienting that direction doesn't need new tech. It needs a shift in perspective. It requires a return to the questions that set purpose, and not the process.
What does the system enable? How do outcomes change? Does it grow human capacity or just activity? Does it bring people closer or pull them into abstraction?

These are not technical limits. They are intentional choices. They decide if the system lives up to its potential or just keeps running without it.

The message was never really about crypto as a label. It was about alignment. Between tech and purpose. Between systems and human outcomes. Between innovation and meaning. When alignment slips away, the system doesn't collapse. It keeps going. But the reason it exists becomes buried under all the motion.

Motion can be mistaken for progress. A system can keep moving—more transactions, more users, more capital—and still not head toward its purpose. Motion looks like forward movement. But without direction, it's just going in circles. The difference between motion and progress is not speed. It's alignment. And alignment does not emerge just from showing up. It has to be named, reinforced, and cared for.

This is where participation versus contribution matters. Participation is easy. It means being there, paying attention, and often depositing capital. Contribution is not the same. It needs intent. It needs an understanding of how your actions fit into a bigger system. Whether what you do pushes the system toward its purpose or pulls it away.

Markets reward participation first because it fuels activity and makes things visible. But contribution decides whether the system produces anything meaningful. When participation rules without contribution, systems become self-referential. They tune for their own activity rather than the problems they were meant to solve. Metrics go inward with a focus on growth, volume, and engagement, rather than outward with importance placed on impact, access, and coordination.

Once that shift happens, it's harder to bring the system back to its original aim. The measures of success no longer point to the original purpose.

There is also a subtle psychological change. If people enter a system through participation rails such as price, speculation, or the lure of returns, they anchor their view of the system to that entry. The first interaction shapes what they think the system is for. If you see it as a transaction, you see it as transactional. If you view it as a bet, you see it as speculative. Over time, that view reinforces itself because people look for information that fits their initial take.

That's how a system's story becomes self-sustaining even when it drifts from its intention. And that's why the original message doesn't just disappear. It gets harder to spot. It sits in the architecture and shows up in what the system can do. But it's not the primary frame people use anymore.

Seek intending to find it. Not reactive attention, quite the opposite: intentional attention. Not focused on what is happening. Focus on what that happening will mean for what the system should do.

Crypto didn't lose its potential. It just lost sight of what that potential was for.

Chapter 5—If People Don't Feel Like They Belong, They Won't Enter

Claire Ross-Brown is an English actress, businesswoman, and philanthropist with extensive experience in both the entertainment industry and the business world.

Launching her acting career at 14, Claire quickly made a name for herself with appearances in popular English TV shows such as EastEnders; and in feature films like The Rainbow. Her expertise also spans the financial sector, where for over thirty years she has been a pivotal figure, advising industry leaders including Cashworks, Goldman Sachs, and Credit Suisse on talent management, corporate partnerships, and strategic deal-making.

In 2019, armed with a deep understanding of modern business and financial practices, Claire founded CJ London, a luxury fashion brand known for its exceptional yet accessible "Timeless Classics" collections. Centered on the iconic "little black dress", CJ London is committed to sustainability, using eco-friendly materials and processes to cater to fashion enthusiasts globally.

In the dynamic landscape of blockchain technology, Claire Ross-Brown joined Concordium, a pioneering Layer One blockchain, as an advisor and news anchor in 2021. Her involvement with Concordium positioned her at the vanguard of blockchain innovation, where she used her experience to offer strategic insights and effectively communicate intricate blockchain concepts to a broad audience. Simultaneously, Claire took the role of Executive Director and presenter at Crypto Channel Direct. This weekly program has quickly become a staple in the crypto community, with Claire dissecting the most pressing crypto and blockchain news, offering opinions, and providing informative segments that explain the industry's most complex topics.

Claire's dedication to demystifying the often-opaque world of cryptocurrencies and blockchain technology made her a trusted voice in the community. Her ability to translate the nuances of digital currencies and the underlying technology into accessible and engaging content ensures that viewers are consistently well-informed and ahead of the curve.

Podcast Guest: Claire Ross-Brown
Podcast Date: January 6, 2024, Season 6
Podcast Link:
https://open.spotify.com/episode/4bDdAgvGBNO39WUle8ZE4h?si=z-81pzIpQUefMbnrFYG-0g

****

If people don't feel like they belong, they won't enter.

A system can be perfectly designed and still fail. Not because it doesn't work. But because people don't adopt it. This layer is the easiest to miss and the hardest to fix. It's not in code or infrastructure or even incentives. It's in perception. In how a system feels the first time someone sees it. In whether they think they can understand it, take part, or add something useful. And that belief—or the lack of it—changes everything that comes next.

Crypto, by design, is supposed to be permissionless. It should remove barriers and let anyone with access join without asking a central authority. But permissionless systems need something less visible and more fragile too: Psychological permission. The inner sense that you belong, that you're allowed to engage, that you won't be exposed for not knowing enough at the start. Without that, openness doesn't feel like opportunity. It feels like intimidation.

Claire Ross-Brown's story starts by knocking down the entire idea that there is one "correct" path into this space. No linear progression. No tidy trajectory. No sequence of credentials that proves you're ready.

Her background, by her own telling, was not just nontraditional; it was fragmented, varied, and "not logical at all." And yet, that lack of a neat path isn't a liability. It's an asset.

If entry into crypto requires a specific background, the system betrays its own premise. It becomes selectively open instead of open to everyone.

What Claire shows is that the early barrier is rarely technical. It's interpretive. People don't first meet complexity. They meet doubt about themselves. They don't ask "How does this work?" at first. They ask, "Am I the kind of person who can get this?" If the answer is no, or even unclear, they stop before they begin.

Most systems fail here without realizing it. They assume openness is enough. That if access exists, participation will follow. But access without confidence is inert. It's there, but unused. Claire's early move was different. She bypassed permission.

At eighteen, she was told she didn't qualify. Too young. No degree. No formal credentials. The system, as defined, had no place for her. She ignored the definition. She walked in. Sat down. Started working anyway. Not because anyone gave her access, but because she assumed it. That assumption matters.

The gap between exclusion and participation is often not structural. It's psychological. The system didn't bend for her. Claire changed how she approached it. And because of that, she entered. Same with crypto. There was no early mastery. No instant comprehension. Like lots of newcomers, she didn't get it at first. But unlike many, she didn't treat that as the end. She treated it as the start.

"There were no silly questions," she asserted. Simple line. But it carries a heavy weight. Learning requires vulnerability. Saying you don't know something in an environment that often rewards acting like you know, especially in fast spaces, feels risky. People don't ask basic things because they assume others already know. They stay quiet, not because they can't learn, but because they worry about how they'll be seen.

"Many people didn't understand it... and were too shy to admit it." That becomes an invisible barrier. The system looks open, but entering it feels closed. Over time, that feeling feeds on itself. The more people think they're behind, the less they join. The less they join, the further behind they seem. Not in skill, but in confidence.

What fills that empty void is speculation. When people enter without a base, they often come in through the loudest door, which is price. It's immediate, measurable, and always moving. It lets you join without understanding. Some of the most followed crypto influencers on social media know very little about technology but can easily shout-talk a story around the price of Bitcoin. But that changes how people participate. They react instead of learning. They follow momentum instead of building conviction. They ask where it's going instead of how it works.

This isn't an individual failure. It's a result of how the system presents itself. Claire's advice cuts through that. "Only invest in what you can afford to lose." "Do your own due diligence." They sound like clichés. Here they mean something else. Not just risk rules. Responsibility. Turning the participant from passive to active. From someone who reacts to someone who engages on purpose. Due diligence, simply put, is understanding before you act. And understanding takes time.

These clichés bring us back to tension. In a world that rewards speed, understanding becomes optional. And when understanding is optional, participation stays shallow. There's another layer that people miss. Access to education is uneven. "Education is a privilege...not everyone has access to it." That matters because crypto is supposed to expand opportunity everywhere. To let people join no matter where they are. But if the ability to understand is uneven, outcomes follow that unevenness. People with knowledge, networks, and time move faster. Others stay at the edge, reacting more and understanding less.

If that gap keeps going, the system repeats the same inequalities it claimed it would fix. Not by shutting people out, but by creating asymmetry. This is where Claire adds something missing from technical talks: Belief. "You can do whatever you put your mind to," she explained. It reads like encouragement. But within the context of accessibility, it's a mechanism.

Belief decides whether someone even tries. Whether they ask the question. Whether they stick around long enough to learn. Whether they move from participant to contributor. Without belief, the system sits unused. Here's the central tension: crypto is meant to be open. But being open doesn't mean people will take part. Participation needs more than access. It needs confidence. It needs understanding. It needs belief. And those things aren't built into the code. They're felt while using the system.

Seen this way, failure isn't the design. It's how people first meet the system. The entry point is shaped by what we talked about earlier: attention, speculation, and speed. People don't come in through purpose, structure, or the chance to change systems. They come in through volatility, stories, and the promise of returns. That first impression matters. It anchors how the system is seen.

Enter through the price and you will see a market. Enter through education, and you'll see a tool. Enter through building and you will see the infrastructure. The system is the same. What changes is perception. Hence, psychological permission matters tremendously.

A system can strip external barriers and still feel closed if inner barriers stay. If people think they're not qualified, not informed, or not capable, they won't act in ways that match the system's potential. They hover at the edges, doing small things, never stepping into the roles the system could enable. Over time, the system shows limited participation. Active, but not fully utilized. Accessible, but not fully understood. Open, but not fully inclusive.

The pattern runs through the chapters. The system was broken. The purpose was set. The tools are there. But the entry point is off.

When entry is off, everything that follows gets turned off too. Exclusion here is rarely loud. No gates. No authority is shutting doors. Technically, the system is open. But perception follows experience, not technical facts. If entering feels confusing, intimidating, or like you don't belong, that feels the same as being excluded. Even if no one stopped you.

That's why psychological barriers are hard to fix. They don't announce themselves. They show up as hesitation, silence, and disengagement. People don't say they're excluded. They just don't take part. They watch. They wait for clarity that never comes. And while they wait, the system moves on without them, making the gap worse. Over time, you get a split in participation. Not by rank, but by familiarity. Those who know the language, the rhythms, the assumptions move easily. They ask questions, test ideas, take risks, and build. Those who don't stay outside. Instead, they use shallow entry points that teach little. They listen to social media influencers who know very little about technology, too.

I only mention that again to drive home this message:

Price, headlines, and surface narratives become stand-ins for understanding. Not because people prefer them, but because they're what's available.

That change nudges the system too. It optimizes for people who are already in. Communication gets inward. Language gets denser. Assumptions about baseline knowledge creep up. Not deliberate exclusion. Just a system maturing around its most active users. The result is the same. The gap between inside and outside widens. And when that gap grows, universal access falls apart.

Access isn't just being able to enter. It's being able to engage once inside. To understand, contribute, and move around without translating every idea from scratch. Without that, participation is superficial. It's there, but it doesn't turn into contribution. This is where inclusion and exposure diverge. Exposure puts people near a system. Inclusion lets them work inside it. The difference isn't technology; it's experience. Does the system meet people where they are or expect them to change completely on their own?

Right now, crypto often favors exposure. It's visible, active, and accessible in form. But visibility is not the same as accessibility. Accessibility is not the same as inclusion. Without deliberate paths that take people from watching to understanding, and from understanding to contributing, opportunity remains uneven. That unevenness isn't obvious at first. People are still entering. They still transact. They still engage at the level they can. But the depth of that engagement varies a lot. And that variation decides who benefits most.

Therefore, the entry problem becomes more than onboarding. It becomes direction. How people enter shapes how they use the system. How they use it shapes what it becomes. If the entry is shallow, the usage is shallow. If the entry is reactive, usage is reactive. If there are enough people at that level, the entire system reflects it.

A system built for everyone fails the moment people feel like it's not for them.

Chapter 6—Open Systems Still Need Boundaries

Anjali Young is the co-founder of Collab.Land.

Anjali has been active in online communities since 1993 as a member and in leadership positions. She has also worked as a lawyer, an adjunct professor, and an early tech startup employee. Besides her interest in online communities and web3 onboarding, she has a passion for NFTs and the artists who create them. While developing her business skills and working on several passion projects, Anjali also homeschooled her two children. Her role at Collab.Land focuses on onboarding many tokenized projects and partners onto the platform, ensuring Collab.Land's "community of communities" is a safe and welcoming space, and educating others about web3 ethos and values.

When she's not working, Anjali enjoys creating art from found objects and improvisational theater. Anjali is one of only a handful of Web3 leaders to have received recognition from Salesforce and was listed as one of its Web3 Advisory Board Members.

Podcast Guest: Anjali Young
Podcast Date: January 15, 2024, Season 6
Podcast Link:
https://open.spotify.com/episode/1fiWllaP9tgz0Ny
RAXdl61?si=2tJ83gGHRkOaR43SFLmwiA

****

Open systems still need boundaries.

A system can be open-sourced, accessible, and even designed to be aligned. And it can still fail. Not because the technology doesn't work. Not because the intention wasn't clear. But because what happens inside a system isn't decided by just the architecture. It's shaped by the people who join and by the behaviors and norms that grow out of how they interact.

When behaviors don't match the system's purpose, the system doesn't fix itself. It reflects them. This is where decentralization shows limits. Decentralization removes central points of control. It spreads authority. It cuts reliance on intermediaries. But it doesn't remove the need for structure. It doesn't replace leadership. And it doesn't guarantee constructive outcomes just because participation is open.

Anjali Young's view comes from something older than crypto: online communities. Long before tokens, NFTs, or decentralized governance, she was in the spaces where people gathered and built relationships without being in the same place. Bulletin boards, early forums, LiveJournal. Not blockchain, but the same basic dynamic. People organizing themselves in shared spaces.

"My life has been very much informed by online communities, " explains Anjali. What those environments showed over time is that openness doesn't make things peaceful. It makes things exposed. People bring their perspectives, incentives, frustrations, and expectations. They don't magically line up because the space is open. They clash, reinforce each other, and change the tone of the place in ways you can't predict. Crypto inherits that. And it makes it louder.

Unlike earlier online groups, crypto communities are both economic and social. Participation isn't just talking. It's about value. People are invested, not just present. That changes how they act. It adds incentives that pull behavior toward ownership, speculation, and often extraction.

From a distance, crypto can look like it has solved community problems. Thousands show up on Discord, Telegram, and other platforms around projects. Tokens help with coordination. Ownership can bring a sense of belonging. Activity is constant and visible across the globe. But presence is not alignment. Ownership is not a shared purpose. A bunch of people holding the same asset are not necessarily a community. It's a set of participants with overlapping exposure.

Whether that exposure becomes something cohesive depends on what happens after they arrive. Left alone, open systems drift toward the lowest common denominator of behavior. "Digital communities are full of trolls." This isn't a moral judgment on people. It's a pattern. Open spaces with uneven incentives make some behaviors louder. Some people build. Others look for chances to take. Some are constructive. Others are opportunistic or extractive. And if there's no way to separate or reinforce those behaviors, the system shows all of them at once.

People act out online often because they feel powerless in real life. That adds a psychological layer people forget in technical talks. Nobody comes in neutral. Everyone has context: personal, economic, or emotional. And when identity is fuzzy and accountability is thin, those contexts show up and become amplified. Not chaos in a movie sense. Just noise. Constant overlapping signals. Hard to tell what's real contribution and what's distraction.

Incentives shape behavior. Crypto adds money to that mix. Participation carries economic possibilities. Tokens mean potential gains. Where there's gain, behavior shifts toward it. That's not automatically bad. It's just how systems work.

The chance of economic gains matters. If the main pull is short-term price moves, people chase visibility and momentum and whatever narrative drives attention. Quick in, quick out. If the pull is long-term value creation, people act differently. They favor stability, steady contributions, and alignment with the project's goals. The system doesn't pick. Participants do.

This is where token gating comes in. Not as exclusion, but as alignment. "If you are putting your money into something, it is up to the community to guide the tone," Anjali explains. Token gating adds friction. It asks for commitment to enter. That changes who shows up. It filters some behaviors and supports others. It doesn't guarantee quality, but it sets a baseline of investment that can help. But even that isn't enough. Structure without leadership is incomplete.

Leadership is not optional. There's a common belief with decentralization that systems will self-organize and that participants will naturally line up behind excellent outcomes if incentives are right. That rarely happens.

"Leaders and founders have an enormous responsibility." Leadership here isn't control. It's direction. It means setting expectations, creating norms, and reinforcing behaviors that match the system's purpose. It means being present, consistent, and accountable to make things stable in a fluid environment. Without that presence, communities don't stay neutral. They decay.

Anjali puts it in terms of give and take. "Commit to your community first; then you will get commitment back." Not a simple trade. A structural thing. Communities mirror the investment they receive. If founders vanish, if communication is spotty, if expectations are fuzzy, people change how much they put in. They pull back. Or they shift to actions that serve themselves over the group.

Commitment is more than a signal. It stabilizes the place.

The tension between short-term incentives and long-term health shows up most in how value gets split. Removing royalties makes communities unsustainable. This isn't just a technical tweak. It sends a cultural message. If systems remove mechanisms that support creators to chase liquidity or lower friction, value moves to traders, not builders. Short-term actions get rewarded over long-term craft. Over time, that alters who sticks around.

The shift from alignment to misalignment is slow. It creeps. Those who are early may still share a purpose. Builders keep building. But growth brings new entrants with different aims and horizons. If those differences aren't woven into a coherent culture, they pull the system apart.

Scale makes it harder. Growth is treated as a win, but growth without alignment just increases variance. More people mean more kinds of behavior. Communication fragments. Norms loosen. The original signal must compete with more noise. The system doesn't stay neutral. It leans toward behaviors that are easiest to maintain at scale. And those are rarely the most constructive. They're the most immediate and the most visible, but are the least tied to shared understanding.

Speculation spreads faster than coordination. Reactions scale more easily than contributions. Short-term incentives move quicker than long-term commitments. That creates cultural gravity. The system is pulled toward the state that takes the least work to maintain. Without counterweights such as clear leadership, reinforced norms, and incentives that align with purpose, that state wins. Not because it's best. But because it's stable given scale.

People adapt. Builders focused on the long term find themselves in places that reward quick gains. Contributors who want alignment run into friction as coherence fades. Some leave, some change, some try to rebuild inside the noise. But without active effort, the system slides to what's easiest, not what's best. Responsibility becomes unavoidable.

In centralized systems, responsibility is concentrated. You can point to who enforces what. In decentralized systems, that clarity is gone. Responsibility is spread out. Distribution doesn't mean people will take it. It can make ownership so diffuse that no one acts. When everyone is responsible, it's easy for no one to move. And the system doesn't stop. Culture forms all the time, through every interaction, every decision, and every signal that's reinforced or ignored.

Once the culture is in place, it's hard to change because new people adapt to what they find, not to what could be different. That's why early stages matter so much. Norms set when a system is small stick as it grows. Not because of strict enforcement, but because they become part of how people see the space. If those norms favor contribution, alignment, and long-term thinking, the system has a foundation. If they favor speculation and short-term gain, those behaviors scale with it. Fixing scale misalignment is a lot harder than avoiding it.

Anjali still cares about inclusion. "Let's keep opportunities open for people all over the world." But inclusion doesn't happen by itself. It needs structure, intention, and ongoing work. You balance openness with boundaries, accessibility with alignment, participation with accountability. Without that balance, inclusion is patchy. And patchy inclusion makes outcomes uneven.

There's also a misconception that removing central control removes the need for coordination. If authority is spread, alignment will magically appear. It won't. Alignment comes from shared understanding, built over time through consistent behavior. It doesn't scale by itself. It needs translation, repetition, and examples. Without that, people default to what they know. They bring market habits, competitive moves, and short-term thinking into a system built for something else. Not out of malice. Out of adaptation. They use the tools they have. But when those tools clash with the system's goals, outcomes follow the tools.

That's why culture and design are inseparable. Culture isn't something you tack on. It's part of the system. It shapes how rules are read, how incentives are ranked, and how people behave. Two systems with the same architecture can end up very different depending on the culture inside them. And culture can't be deployed like code and forgotten. It's kept alive by people taking part. It's reinforced by behavior. It's held up by accountability.

Without accountability, it doesn't remain neutral. It drifts to what's easiest given the system's conditions. The last layer of the problem isn't whether the system functions or whether people can access it. It's whether the environment people enter supports the behavior the system was supposed to enable. If it doesn't, the system still runs. It just won't do what it was meant to.

By now, the pattern across the chapters is clear. The system was broken. The purpose was stated. The tools exist. People can join. But without culture, boundaries, leadership, and aligned incentives, the system does not hold. It fragments.

Decentralization doesn't remove responsibility. It redistributes it. And if no one picks it up, everything falls apart.

Chapter 7—Finding the Message Again

There's a difference between lost and forgotten.

Lost means out of reach. Out of memory. Out of recovery. Forgotten means it's still there. Exactly as it was. Waiting under the surface. Not gone. Just not what people are looking for.

Crypto didn't lose its message. It forgot it. The signal is still there. Underneath the movement. Underneath the noise. Underneath how attention shapes the space from the outside. The same problems that started this are still here. The same urgency is still there. Control over money is still concentrated. Access is still uneven. Coordination is still clumsy. The systems that make this happen mostly stay hidden until they don't.

Nothing fundamentally changed. So nothing fundamental was disproven. What changed was the focus. Read the conversations in this book and a pattern shows up. Not a theory. A structure. The system was broken. The purpose was apparent. The tools were available. People could join. Communities could form. And yet something didn't hold.

What links those layers is not technology. It's attention. Attention decides what gets built, what gets funded, and what survives. And when attention keeps moving and is pulled by what's new, by what's visible, and by what needs to be explained right now, the direction of development follows it.

This isn't abstract. You can see it. When attention is on price, capital moves to price-sensitive things. When attention is on narrative, projects tune for narrative. When attention is on speed, developers favor what scales fast over what should be done right. This is not corruption. It is alignment. Builders follow incentives. People follow visibility. Systems change based on what gets rewarded.

Over time, those rewards point the way. Price becomes the main signal because it's immediate. It moves. It updates. It gives quick feedback. It lets people read the system without really understanding it. That matters a lot because understanding takes time. Interpretation does not. And when time is short, interpretation wins.

Purpose doesn't move that fast. Systems don't update instantly. Meaning doesn't swing on a chart. Thus, they get less attention. You get a split between what's easy to see and what's hard to grasp. A split between the immediate and the structural. Between what happens and what it actually means. In the long-term, that split shapes the space.

The message didn't vanish. It sank into the background. To get it back, you don't invent something new. You notice what's already there. The system's structure hasn't changed under the surface.

Dr. Karl Huber points to the core issue. Control of money stays concentrated. Centralized actors still influence access, tweak supply, and decide who joins. Market cycles and new layers haven't fixed that. These are structural facts that persist no matter how the surface changes.

Violet Abtahi points to the problem of purpose. That's still unresolved. Moving beyond survival. Coordinating resources to unlock human potential. Building systems that are more than efficient, that are transformative. Those outcomes aren't at scale. Tools exist, but the direction hasn't been steady. Systems were built without always being clear about what they should produce.

Claire Ross-Brown points to access. That gap is still real. Technical openness is not the same as psychological openness. People can get in. But they don't always feel able to take part. Without understanding, without confidence, without belonging, access is incomplete. The system is available but not fully utilized.

And Anjali Young points to culture. Open systems mirror participant behavior. Without boundaries, leadership, or aligned incentives, communities drift. They don't fail fast. They decay slowly. And that decay shapes what the system becomes.

These are not separate problems. They're layers of the same thing. Fix the system, but ignore purpose, and direction is lost. Say you've defined a purpose but don't enable access, and people can't take part. Open access, but forget culture, and systems fragment. Shape culture but ignore system limits, and constraints stay. Each layer depends on the others. Pull one out and the structure weakens. Skip one, and the outcome changes. That's why the message matters. It's not a slogan. It's a structure.

You need:

1.   A system that supports individual control.

2.	A purpose that goes beyond survival.

3.	An entry that lets people participate.

4.	A culture that keeps things aligned.

Without all four, the system keeps running. But it produces something else. The system doesn't fail.

It drifts.

Drift is sneaky. It doesn't shout. It looks like continuity. Transactions keep happening. New people still join. Capital still flows. From the outside, nothing looks broken. So the urgency feels low.

Drift quietly redirects. Slowly. No mass agreement is needed. Thousands of tiny choices add up and momentum builds. Momentum isn't often questioned. It gets followed. But momentum is not the same as correctness. It only points in a direction. And unchecked directions become an outcome.

That's where individuals matter. In decentralized setups, there is no central course correction. No single force to restore intent. Only participants. And they act on what they perceive. If perception is partial, behavior is partial. If perception is off, the system becomes off. This isn't a flaw in decentralization. It's what it does. It removes control and spreads responsibility. Across everywhere and everyone.

This creates a paradox. The system lowers dependence on central authority. But it raises dependence on individual awareness. Without that awareness, the system doesn't default to its original aim. It defaults to whatever behavior gets reinforced. And attention shapes those behaviors. The conversation moves from design to human action.

The system enforces rules and keeps transactions honest. It doesn't create meaning. People do that. The same system can lead to extraordinarily different results. Not because technology changes, but because behavior changes. Behavior comes from perception.

Back to attention again, what people focus on shapes how they read the system. How they read it shapes how they act. How they act shapes what the system becomes. That's the feedback loop. Not technical. Behavioral. Once it starts, it feeds itself. People who use crypto like a market reinforce market moves. People who use it like a system reinforce structural work. Both can happen at once. They don't lead to the same results. The system doesn't pick a path. People do. Eventually, the dominant behavior wins out.

Reclaiming the message isn't about going back. It's about seeing what's still under what has changed. The system grew. Tools got better. Infrastructure improved. But the core idea wasn't replaced. It was just covered up. And being covered up isn't forever. It lasts only while attention is elsewhere.

That means it can be flipped. Not by force. Not by grand coordination. By recognition. Individuals can choose to act differently. They can choose to look past the obvious. And questioning what seems to matter. See what matters most.

This isn't a call for everyone to agree. It's a call to pay attention. Attention changes behavior. Behavior changes systems. Not overnight. But for sure. Direction shifts the same way drift formed. By accumulation. The same slow process. The difference is intention.

Intention starts with attention. Failure gets noticed. It pulls attention. It forces action. Drift doesn't. Drift feels like progress. Growth, users, activity. All the signs. It doesn't get questioned. But drift has a direction, and that direction decides the outcome. This moment matters because the gap between perception and structure is big enough to matter.

The system still works. But what it makes is drifting from its original reason. That creates tension between what it does and what it should do. Technology doesn't fix that. Attention does. Reclaiming the message needs a shift. Not in tools. Not in protocols. In how people engage. From reaction to intention. From participation to contribution. From watching to understanding.

This shift is not automatic. It's personal. It happens one person at a time. In what they focus on. In how they make sense of what they see. In what they decide is important. Systems mirror behavior. If people treat crypto like a market, it becomes a market. If they treat it like a system, it becomes a system. The tech doesn't decide. Choice does.

There will be cycles. Volatility. Noise. Those are features, not bugs. Beneath them is something steadier. The original idea. People control their own assets. Access not decided by central power. Economic value moves without permission.

Opportunity is not limited by place, rank, or proximity to power. That idea is still there. Just overshadowed.

And what's shadowed isn't gone. It's waiting. Waiting for attention to come back. Waiting for the intention to follow. Waiting for alignment to rebuild what drift pulled apart. The message doesn't need rediscovery. It needs remembering.

Remembering isn't passive. It's active. It means engaging with the system at the level it really works. Looking past the easiest signals. Questioning what looks important. Understanding what actually matters. This isn't rejecting what the system has become. It's a rebalancing. A reminder that what's visible isn't always what's fundamental. And what's fundamental doesn't disappear just because it's out of focus.

What makes now different is not clearer message. It's that ignoring the message now costs more. As the system grows, the results of its direction matter more. And direction is not set by technology. It's set by attention.

The message was never lost. It was simply waiting for people to listen again.

Appendix A: About the Author

Jamil Hasan is the founder of Crypto Hipster Publications and host of the Crypto Hipster Podcast, a platform dedicated to one core idea: where builders talk freedom, not price.

Across over 580 interviews, he has focused on founders, entrepreneurs, and independent creators—the people actively building the digital economy from the ground up. His work is not about market cycles or short-term narratives, but about understanding the deeper motivations, risks, and convictions behind those creating new systems.

Jamil's approach is rooted in filtering signals from noise. In an industry often driven by speculation and surface-level commentary, he has built a platform that prioritizes long-term thinking, first-principles discussion, and authentic perspectives.

Before fully committing to the digital asset space, Jamil built his career across financial services and data-driven initiatives, including early-work connected to blockchain, artificial intelligence, and client-facing strategy. That foundation informs his ability to translate complex technological change into clear, human insights.

After stepping away from his work during a period of intensive medical treatment, he returned with a sharper lens and a more defined mission: to elevate real builders, challenge conventional narratives, and document the ideas shaping the future of the digital economy.

He returned with a sharper lens and a more defined mission: to elevate real builders, reject noise disguised as insight, and focus only on what endures.

Signals Through the Noise represents the next evolution of that mission.

Appendix B: Crypto Hipster's Curtain Calls Episodes

A complete list of Jamil's Crypto Hipster's Curtain Calls podcast compilations from Seasons 6-8 is presented below. These compilations set the basis and foundational groundwork for the current Crypto Hipster's book series. Each podcast episode, except for the first one, starts with a question, and each book answers that question. These episodes comprise fifty guests from Season 6, one hundred guests from Season 7, and one hundred thirty-one guests from Season 8.

All of them can be found at anchor.fm/crypto-hipster-podcast and can be listened to on Spotify, Apple Podcasts, Amazon, YouTube, Anchor, or wherever enjoy your favorite podcasts, including the full interviews from each of the guests presented in this book.

Ep. 1: Insights On Leadership, Sustainability, and the Global Digital Economy

Ep. 2: The Lost Message? How Crypto Is Still About Financial Freedom and Global Socio-Economic Betterment.

Ep. 3: Bitcoin: A Beauty or a Beast? Why Bitcoin and All Its Features Will Stand the Test of Time.

Ep. 4: FTX: The First of Many Crypto Implosions to Come? How We Can Overcome Additional Existential Challenges in Web3.

Ep. 5: People are People, so Why Should It Be? Intellectual Property Advancement Should Develop So Effortlessly in Web3.

Ep. 6: Sam's Coin? How Solana's Continued Development Has Strengthened Its Network and Transformed It Into Much More Than Just "Sam's Coin."

Ep. 7: A Dopamine Kick? How to Avoid Getting Obliterated When Chasing 10,000X.

Ep. 8: The Ghost in You? Seeing Yourself Through the Eyes of Decentralized AI.

Ep. 9: Our Independence Day? How Crypto Can Improve Our Daily Lives and Create Paths for Personal Freedom.

Ep. 10: Mercy, Mercy Me? Why We Should Advance Sustainability Initiatives Using Generational Decentralized Technology.

Ep. 11: Our Lips Are Sealed? Why We Should Think and Talk About a Future Filled With Helpful AI Agents.

Ep. 12: Gods of War? How to Navigate the Bitcoin Cycle Waves Proactively Without Running Riot.

Ep. 13: Peace of Mind? How to Break Free From Instant Dopamine Kicks and Make Solid, Long-Term Crypto Investing Choices.

Ep. 14: Land of Confusion? Helping Investors and Traders Navigate the Wild West of Cryptocurrencies.

Ep. 15: Bizarre Love Triangle? A Web3 Bill of Rights, Crypto's Good Ole Days, and the Three Magi of Blockchain.

Ep. 16: Another Day in Paradise? Why The OG Cryptocurrencies Need Saving and How a Bright Future Ought to Become Reality.

Ep. 17: Stairway to Heaven? Why We Are Climbing A Destined Pathway Toward a Fully Tokenized Future.

Ep. 18: More Than a Feeling? Re-imagining Sports Betting Through Decentralization.

Ep. 19: I Like to Move It, Move It? Sparking a Rapid Explosion of the Decentralized Entertainment Industry.

Ep. 20: Money for Nothing? Understanding the Future of Digital Currencies Beneath the Hype Hood.

Ep. 21: Groove Is in the Heart? How to Get Your Groove On by Using Artificial Intelligence Agents.

Ep. 22: Fields of Gold? How the Future of Bitcoin Shines Brightly Even When the Price Sometimes Drops to a Number Unsightly.

Ep. 23: Little Pink Houses for You and Me? How Independent Creators Can Trust the Crypto Industry Again.

Ep. 24: Eye in the Sky? Taking a World View of the Future of Digital Identity, E-Commerce, and Web3 File Storage.

Ep. 25: Wheel in the Sky? Focusing on the Fun Journey of Web3 Gaming.

Ep. 26: Blue Sky Action? How to Navigate the Complexities of Crypto Auditing, Accounting, and Financial Reporting.

Ep. 27: You Down With O.P.P.? Why Revolutionizing Blockchain Security Will Keep Others From Coveting and Stealing Your Property.

Ep. 28: Let's Make Lots of Money? Unlocking the Possibilities of Enriching the World Using Decentralized Finance Platforms and Protocols.

Ep. 29: The Grand Illusion? Why Ushering In a New World of AI Superintelligence is Much More a Reality Than an Illusion.

Ep. 30: Only Time Will Tell? Overcoming Hackers, Pandemics, and Identity Threats Through Unique Cryptographic Solutions.

Ep. 31: Only the Young? How to Make Decentralized Finance (DeFi) a Safe and Attractive Journey For All Generations To Benefit.

Ep. 32: We've Only Just Begun? How We Are Just In the Early Innings of the Global Decentralized Finance Evolution.

Ep. 33: Rain on the Scarecrow? How to Build Web3 Gaming Solutions That Sustain the Original Ethos of the Blockchain Industry & Bypass Bankers' Bloodthirsty Greed to Destroy It All.

Ep. 34: Two Tickets to Paradise? Why the Future of Non-Fungible Tokens is Bright Despite Having Troubles in Paradise.

Ep. 35: Dancing in the Dark? Why Many Crypto Founders Are Fumbling Around In the Dark Regarding Global Crypto Regulations and What They Can Do About It.

Ep. 36: That Was Yesterday? How to Implement Forward-Looking Regulatory Solutions to Some of the Industry's Greatest Challenges.

Ep. 37: I've Got the Power? How Web3 Can Help Content Creators Regain Ownership Over Their Personal Content and Grasp the Power Back From Big Tech Firms That Seek to Exploit Them.

Ep. 38: Ain't We Funkin' Now? How Blockchain Technology, Through Innovation and Education, Are Causing Breakthroughs in Some Funky-Cool Markets.

Ep. 39: In the Name of Love? They Can Try to Take Away Our Crypto, But They Cannot Take Away Our Pride.

Ep. 40: Why Can't This Be Love? Why Passion and Social Impact Remain at the Heart of Web3 Development.

Ep. 41: Don't Come Around Here No More? Why Leveraging Blockchain Technology to Make a Difference for Humanity Continues to and Always Will Matter Most.

Ep. 42: Home By The Sea? Why Decentralized Finance is Primed to Outperform Traditional Finance Even During the Most Turbulent Tides.

Ep. 43: Sultans of Swing? Empowering Crypto Traders and Investors with Useful Tools and Opportunities to Take Full Advantage of Their Web3 Experiences.

Ep. 44: 1979? While Artificial Intelligence is Nothing New, How We Harness Its Power Has Transformed Dramatically In the Last Half-Century.

Ep. 45: Walking on the Moon? Leveraging AI Models to Transform Amorphous Ethical Ideals Into Solid Earthly Business Practices.

Ep. 46: 2000 Light-Years From Home?
Courageously Exploring the Impending Universal
Explosion of Real-World Tokenized Assets.

Ep. 47: Every Rose Has Its Thorn? Despite a
Thorny Path Thus Far, Why the Future of
Blockchain Development is On An Increasingly
Smoother Road Heading Toward Tomorrow.

Ep. 48: Another Brick in the Wall? Why
Blockchains Must Be Implemented Brick by Brick
Globally for Everyone's Benefit.

Ep. 49: Boogie Wonderland? How
Programmable Decentralized Money Will Help Us
All Embark Upon Creating a Self-Sovereign,
Trusted Web3 World.

Ep. 50: Boogie Nights? Learning Life Lessons
and Web3 Trading Strategies From Crypto's Role in
Liquor, Laundering, Lust, and Luxury.

Ep. 51: The Sky is Crying? How Web3 Financial
Engineering Has Altered the Purpose and Vision of
Blockchain Technology From Idealistic Altruism to
Profit Maximization.

Ep. 52: Invisible Touch? How to Fend Off Your
Inner Demons While Trading Cryptocurrencies by
Using Automation Tools, Security Measures, and
Risk Management Techniques.

Ep. 53: A Touch of Grey? How the Power of Blockchain and Web3 Extends to Even the Most Illiquid Markets and Grayest Areas of Investing.

Ep. 54: Out of Touch? Why Transforming On-Chain Compute Into Real World Tangible Assets is Not Such a Remote and Distant Possibility.

Ep. 55: Thunderstruck? How Zero-Knowledge Strikes Open the Chasm Separating a New, Abundant Decentralized Future from the Ancient and Nearly Extinct Traditional Financial System.

Ep. 56: Welcome to the Jungle? How to Navigate What Lurks in the Jungle of Decentralized Finance Successfully.

Ep. 57: Wanted Dead or Alive? Why We Should Stop Listening to Critics Who Falsely Pronounce Bitcoin's Death, and Focus Instead on Building an Open, Collaborative, Better Society.

Ep. 58: Mainstreet? How Wall Street's "Bitcoin Capture" Robbed Main Street Investors and What We Can Do About It.

Ep. 59: Rock of Ages? Why Crypto Has Been Resilient Through Multiple Storms Despite Threats From Every Direction and Evil Forces Trying to Destroy It.

Ep. 60: Peaceful Easy Feeling? How to Overcome Systemic Challenges So Traders and

Investors Can Confidently and Serenely Navigate the Crypto Market.

Ep. 61: Shiny Happy People? How Crypto Has Created a New Class of Economic Survivors and Affluent Outcasts From the Traditional Financial System and Social Networks.

Ep. 62: Heaven's On Fire? How Decentralized Infrastructure's Evolution Sparked the Inferno to Burn Bridges to the Ground and Build an Inclusive World for Financial Freedom.

Ep. 63: Fire and Rain? Unifying Fragmented Economies to Introduce the Next Billion Users to the Web3 World.

Ep. 64: We Didn't Start the Fire? Why Blockchain's Evolution Is More Important Than Greedy Bankers & Crooked Politicians Making a Quick Buck for Themselves at Your Expense.

Ep. 65: Edge of Seventeen? How the Failures and Successes of the 2017 Initial Coin Offering Era Led to Later Advancements in Cryptographic Innovation.

Ep. 66: Bitter Sweet Symphony? How Enduring the Growing Pains of a Nascent Web3 Industry Today Will Lead to a Thriving Global Economy Tomorrow.

Ep. 67 (FINALE): Jump? Taking the Leap Out From the Pains of the Old World and Into the Joys of the Web3 Future.

Appendix C: Crypto Hipster Podcasts

A complete list of Jamil's Crypto Hipster Podcasts from Seasons 6-8 is presented below. All of them can be found at anchor.fm/crypto-hipster-podcast and can be listened to on Spotify, Apple Podcasts, Amazon, YouTube, Anchor, or wherever enjoy your favorite podcasts, including the full interviews from each of the guests presented in this book.

This appendix lists Seasons 6, 7, and 8 podcasts by category and genre. For Seasons 1 through 5, please check out some of my Crypto Hipster Chronicles and Crypto Hipster Mysticals books and compilations, or listen to my podcasts at Crypto Hipster wherever you listen to your favorite podcasts.

Accounting and Finance for Cryptocurrencies

- How to Best Navigate the Complexities of Crypto Accounting, with Rich Zhou @ Aquifer CFO

- Building a CPA Firm That Not Only Understands Digital Assets But Also Serves as a Trustworthy Strategic Partner for Web3 Businesses, with Patrick Camuso @ Camuso CPA

- Creating Consistency, Consensus, and Auditing Standards Across the World of Blockchain Technology, with Hind Kurhan @ Thesis* Defense

- Designing Trusted Environments to Drive Mass Adoption of On-Chain Finance, with Jeff Owens @ Haven1

Creating Artificial Intelligence Agents

- Building a Decentralized Platform Where Anyone Can Create Their Own Custom AI Agents, with Colin Fitzpatrick @ Griffin AI

- Exploring the Future of AI Agents and Their Applications for Decentralized Finance and Banking, with Bullet Bulat @ ReDeFi

- Pioneering the Creation of Emotionally Intelligent AI Companions and the First IP-Powered AI Protocol, with Max Giammario @ Kindred

- Building an On-Chain Arena to Reward the Best AI Agents, with Andrew Hill @ Recall

- How to Design Decentralized Infrastructures That Make Artificial Intelligence Agents Reliable, with Karan Sirdesai @ Mira Network

- Making AI Accessible Through Community-Driven Data Training, with Johanna Rose Cabildo @ Data Guardians Network (D-GN)

- Tapping Into a Global Contributor Network to Source AI Verified Insights, with Rowan Stone @ Sapien

- Movement in Action!: How Personal Customized AI Agents Can Help Improve

Your Health and Catapult the Movement
Economy, with Oleg Fomenko @ SWEAT

AI Modeling

- One Hundred Years of Artificial Intelligence and Its Future with the Intersection of Blockchain Technology, with Dr. Alok Aggarwal @ Scry AI

- Why We Should Create a Collaborative AI-Powered Blockchain Economy That Benefits Everyone, with Sean Ren @ Sahara Labs

- Discovering the Dolcelorian, an AI Agent Uniquely Suited for a Platform Rebellion, with Justin Banon @ Boson Metasystem

- Crypto Hipster Podcast Episode 300. A Vision for Change through the Eyes of AI, with Paulius Stankevicius

- 2024 Holiday Special: Why Preventing Theft and Fraud By Using AI in Crypto Matters, with Michal "Mehow" Pospieszalski @ MatterFi

- Ushering in a New World of Sovereign Superintelligence, with Ahmad Shadid @ O.xyz

- Building AI 3.0 to Overcome Systemic Challenges Within the Web3 Industry, with Chen Feng @ Autonomys

- Exploring the Intersection of Human-Generated Data, Blockchain, and Artificial Intelligence, with Kurt Ivy @ HumanizedAI

- How to Architect a Decentralized Global Data Marketplace for the AI Era, with Brendan Playford @ Masa

- Creating Scalable Infrastructures and Decentralized Systems That Reward AI Model Creators, with Erick Ho @ Function Network

- Harnessing the Power of Artificial Intelligence and Machine Learning to Change the Game and Transform the World, with Wei Xie @ ArenaX Labs

- Crypto Hipster Presents: Reporter on the Ground, Episode 10: Pioneering Advancements in Secure, Collaborative AI Model Training and Deployment, with Jiahao Sun @ FLock.io

- How to Harness the Power of High-End GPUs for Scalable and Efficient AI Model Training, with Jakub Ondrášek @ Clore.AI

- How to Democratize 3-D Content Creation With Advancements in AI Architecture, with Ben James @ 404-GEN

- Unifying Access to Top DeFi Protocols That Secures Capital Deployment Into Decentralized AI, with Yaroslav Writtle @ Yelay

- Examining the Importance of Transparency and Verification in AI Decision-Making Processes, with Jason Teutsch @ Truebit

Bitcoin

- How to Make Bitcoin Transactions Safe, Simple, and Accessible for Everyone, with Sung Choi @ Coinme

- Why Bitcoin Is Unbreakable in the Face of Black Swan Global Events, with Agne Linge @ WeFi

- Crypto Hipster Presents: Shooting from the Hip, Episode 1: The Meaning and Beauty of Bitcoin Maximalism, with Vlad Costea @ Bitcoin Takeover Podcast

- Bitcoin & Beyond: Pioneering Innovation in the Crypto Payments Industry with Mark Højgaard @ Coinify

- Why the Next Crypto Winter and a Bitcoin Price Drop to $60K Is Now in the Cards, with Clem Chambers @ aNEWfn.com

- How to Ride the Bitcoin Halving Wave Using Investing Techniques Everyone Should Know, with Edward Mehrez @ Arrow Markets

- Why Bitcoin-Native Xapo Bank Members Are Using Bitcoin as a Retirement Investment Product, with Seamus Rocca @ Xapo Bank

- Ordinals, Optionality, and Taking a Pragmatic Approach to Building Out the Bitcoin Mining Stack, with Sheldon Bennett and Steven Eliscu @ DMG Blockchain Solutions

- How to Build a Tax-Free Retirement Nest Egg With Self-Directed Bitcoin IRAs, with Adam Bergman, Esq. @ IRA Financial

- Crypto Hipster Presents...Shooting from the Hip!, Episode 7: Solving the Misaligned Incentive Strategies with Bitcoin and Ethereum, with David Lancashire @ Saito

- How Bitcoin Ordinals Can Pave the Way for Individual "Tokenized" On-Chain Identities and Financial Inclusion Globally, with Taha Abbasi @ Ferrum Network

- Bitcoin 3.0: Developing the Next Evolutionary Phase of the Bitcoin Ecosystem, with Rajiv Khemani @ Auradine

Blockchain Implementation

- Onboarding the Next Wave of Developers into Web3 and Supporting Them to Build Intelligent Infrastructures, with Johnny @ Empyreal SDK

- Why Now Is the Best Time to Build Infrastructure That Fosters the Development of Institutional Financial Systems on the Bitcoin Blockchain Network, with Vakeesan Mahalingam @ Torram

- Pioneering the Future of Digital Asset Management and Blockchain Integration, with Darren Carvalho @ MetaWealth

- Increasing Intelligence, Understanding Money, and Developing a Customer-Centric Approach to Building Your Web3 Business, with Cyrus Taghehchian @ SHOPX

- Why Polkadot's Ambassador Program Will Help Create a Bright Future for Blockchain Developers, with Lucy Coulden @ Polkadot

- Empowering Developers to Build Compliant Web3 Applications That Protect User Privacy, with Joe Andrews @ Aztec Labs

- DAOs, "Dots", and the Future of Web Three Developers, with Bill Laboon @ Web3 Foundation

- Empowering Analysts and Developers to Query, Create, and Build the Future of Finance Using Proprietary Web 3 Data Algorithms, with Jim Myers @ Flipside Crypto

- Why Everyone Should Have Access to On-Chain Payments Across Multiple Blockchains, with Adrien Stern @ Reveel

- Helping Early-Stage Founders Leverage Blockchain Technology to Bring Forth Their Ventures to the World, with Aly Madhavji @ Blockchain Founders Fund

- How Simplifying Blockchain Access Helps Users Build "Whatever, Wherever", with Yair Cleper @ Magma Devs / Lava Network

- Creating a Richer and More Robust Blockchain Developer Space in Nigeria with a 4-Week Master Class, with Jathin Jagannath and Awoskia Israel Ayodeji

- Smart Contracts, Programmable Money, and Building an Interoperable Technology World, with Dr. Weijia Zhang @ Wanchain

- How to Connect the Global Crypto Ecosystem on One Platform, with Peter Kris @ Gasp

- Solving the Siloed Network Environment Problem to Embark Upon Building a Native, Decentralized, Multi-Chain, Web Three World, with Omer Sadika @ dWallet Labs

- Building Discreet Log Contracts To Create Trustless Bridges Between Bitcoin and Ethereum and Foster the Evolution of Decentralized Finance, with Aki Balogh @ DLC.Link

- How to Unify and Enhance Accessibility of Blockchain Data Across Various Networks, with Bunny @ DORA

- Building the World's First Layer-One Programmable Data Chain, with Josh Benaron @ Irys

- Building the Web3 Generation of "Amazon Web Services" through Cloud Service Providers, with Sebastian Pfeiffer @ Impossible Cloud Network

- How to Build an Open, Serverless, and Permissionless Compute Network that Drives Positive Change for a Global Society, with Alison Haire @ Lilypad

- Leveraging Crypto to Dissolve Barriers That Have Relegated the Unbanked and Under-banked to the Financial Sidelines, with Kathy Roberts @ Switch Reward Card

- Creating a New Class of Reactive Decentralized Web3 Applications on a Dream Computer, with John Vibes @ Somnia.Network

- Envisioning a Future Where Banks Embrace Blockchain Technology, with Lindsey Lim @ Radix Foundation

- How Based Roll-ups Can Help Even the Smallest Fish Swim Mightily in the Largest Pond, with Amir Forouzani @ Puffer Finance

- Unlocking the Future of Blockchain with Zero-Knowledge Proofs, with Teemu Päivinen @ ZkCloud

- Pioneering the World's First Super-Computer Powered by Handheld Devices, with Butian Li @ Bless Network

- Crypto Hipster Presents: Reporter on the Ground, Episode 5; Building Cryptographically Secure Cold Storage Facilities for Institutions, with Oliver von Landsberg-Sadie @ MPCH

- Building, Scaling and Developing Large-Scale Distributed Cloud Systems on the Blockchain Ground-Floor in Serbia, with Bogdan Habic @ Tenderly

- How to Empower the Automation of Day-to-Day Operations and Innovate Today Rather Than Months from Now, with William Herkelrath @ K3 Labs

- How to Help Web3 Developers Easily Deploy Hybrid dApps on Telegram, with Pavel Altukhov @ TAC

- Advancing the Technical Frontier and Evolution of the NEAR Protocol, with Bowen Wang @ NEAR One

- How to Create a Thriving Prediction App System On-Chain, with Dan Kaizer @ Azuro

- Crypto Hipster Presents: Reporter on the Ground, Episode 4; Why Disaster Recovery Processes are a Critical Component of Building Crypto Custody Infrastructures, with Haden Patrick @ Cordial Systems

- How to Empower Decentralized Order Book Exchanges with Tim Wang @ Elixir

- Unlocking the Full Potential of Blockchain by Creating the First Universal Layer Two Protocol with Native Restaking, with Karan Bharadwaj @ Arithmic

- Why Decentralized Physical Infrastructure Networks are blockchain's next big trend, with Michael O'Rourke @ Pocket Network

Blockchain Security

- Exploring the Benefits of Revolutionizing Blockchain Security Measures that Protect Us from Key Exploits, with Riad Wahby @ Cubist

- Crypto Hipster Presents…Shooting from the Hip! E14: How Creating Robust Economic Security Helps to Fend Off Even the Strongest Adversaries and Weakest Protocol Security Infrastructures w/ Dan Hughes

- Creating a Universal Web Three Privacy Encryption Layer to Secure the Decentralized World, with Julian Deschler @ Elusiv

- When Does the Bull Run? Crypto and Blockchain Security Insights for the Upcoming Bull Market, with Hexens

- How White Hat Hackers Are Transforming Web Two Cyber Security Initiatives to a Web Three Future, with Sipan Vardanyan and Vahe Karapetyan @ Hexens

- Reimagining Blockchain Security and Multi-Party Computing with an Innovative Approach to Building Digital Asset Networks, with Michael Cunningham @ io.finnet

- How to Create Bespoke Solutions that Solve New Identity Threats Posed by Artificial Intelligence and Bots, with Kitty Horlick @ Rarify Labs (Part 1 of 2)

- Global Pandemics, EMP Threats, Cyber/AI Attacks, Civil War, and How Blockchain and a Survivorship Community Can Help, with Dr. Drew Miller @ Fortitude Ranch

Crypto Trading and Investing

- How to Best Empower Users to Manage Their Non-Custodial Wallets Intuitively, with Zhen Yu Yong @ Web3Auth

- How to Conduct Your Own Digital Asset Health Check and Ensure Your Crypto Investing Decisions are Sound, with Anthony Fernandez @ ICONOMI

- How to Avoid Getting Burned While Undertaking a Digital Security Revolution, with Michal "Mehow" Pospieszalski @ MatterFi

- Why Meme Coins Can Solve the Inherent Inequities of Financial Nihilism, with Rennick Palley @ Stratos

- Galileo FX CEO Shares His Insights On How Anyone Can Become An Expert Crypto Trader, with David Materazzi

- How to Uncover Hidden Stories Behind Crypto Prices and Use Custom Analytics to Solve Blind Spots in Your Investment Thesis, with Aakash Athawasya @ PYOR

- How to Avoid Chasing Greed and Illicit Activity During Crypto Bull Markets, with Phillip Alexeev @ CrossFi

- How to Become a Successful Crypto Trader by Developing a Winning Trading Mindset, with Casey Stubbs @ Global Prop Trader

- Optimizing the Crypto Trading Experience with Perpetual Futures for Bitcoin, Altcoins, and Meme Coins, with Mohd Kifa @ Flipster

- Helping Crypto Traders and Investors Take Advantage of the Best Strategic Opportunities to Generate Yield and Enhance Efficiency Through Liquid Staking, with Michael Wasyl @ Bracket

- Learn, Invest, Succeed—How Knowledge, Education, and Mindset Can Empower Investors and Traders to Navigate Crypto and Commodity Markets, with Mukarram Mawjood @ Bullionite Asset Group

- Crypto Hipster Presents...Shooting from the Hip! (Ep. 13): How to Spot Wash Trading, Rug Pulls, and Various Sordid Unethical Behaviors in Crypto Markets Easily; with Mathias Beke @ Kairon Labs

- Why Public Blockchain Wallets Are Dead and Privacy Matters Most, with Georgi Koreli @ Hinkal

- Making Digital Transactions Simple Through Text-to-Trade Functionality and "Hooting" with Dylan Dewdney @ Kuvi.ai

- Unlocking the Power of Decentralized Finance to Help Investors Stay Safe, Transact Ethically, and Gain Access to Web3, with James Toledano @ Savl

- Building a Secure, Robust, and Intelligent Web3 Wallet of the Future, with Alvin Kan @ Bitget Wallet

- Lessons and Insights from Helping Private Clients Invest in Crypto Successfully, with Marc Walton @ Forex Mentor Pro

- Whiskey, Wine, and Watches: Where Passion and Crypto Meet to Create New Investment Opportunities, with Sam Mudie @ Savea

- Crypto Hipster Presents...Shooting from the Hip! Episode 8: Safeguarding Investors Against Pump and Dump Schemes by Building a Best-in-Breed, Multiple Asset-Backed Cryptocurrency, with Jack McInerney

- The Challenges with Solving Inauthentic Web3 Washing, with Justin Banon @ Boson Protocol

- Building the World's First On-chain Order-book Exchange to Allow Users Full Control of Their Crypto Assets, with Vitali Dervoed @ Spark

- Transforming Family Offices and Innovating Special Purpose Vehicles for the Web3 Digital Age, with Jake Claver @ Digital Ascension Group

- How the UK is Helping Professional Crypto Investors by Enabling Exchange Traded Notes, with Oliver Linch @ Bittrex Global

- Options Trading, Digital Dollars, and Crypto Predictions for 2024, with Anthony Saliba @ Liquid Mercury

- Insights on the Future of Crypto Derivatives Trading, with Mark Lee @ SynFutures

- Why Crypto Options and Structured Products Can Help Manage Risk as We Head into the Bull Market, with Georgii Verbitskii @ TYMIO

- How to Transform Web3 Capital Allocations Through Successful Grant Funding Programs and Strategies, with Meg Lister @ Gitcoin Labs (Video)

- How to Leverage the Benefits of Automation Tools When Investing and Trading for the Next Decentralized Finance (DeFi) Summer, with Chris Bradbury @ Summer.Fi

- Preserving Security, Transparency, and Efficiency in Decentralized Trading Environments, with JOTARO @ JOJO

The Evolution of Decentralization

- Building an Interconnected Network of Value to Bring Back Crypto's "Good Old Days" and Foster a Blockchain Renaissance, with Jerry Li @ Artela Network

- Crypto Hipster's Christmas Special: The Three Magi of Blockchain Technology: Speed, Security, and Scalability, with Dr. Alan Tominey @ Gorki

- How to Transform the Crypto Industry into a $100 Trillion Market Cap Opportunity, with Nick Cowan @ VLRM

- Crypto Hipster Presents: Reporter on the Ground, Episode 3; Why Zero-Knowledge is the Next Really Very Smart Trillion Dollar Opportunity, with Alex Pruden @ Aleo Network Foundation

- The Birth, Evolution, and Future of Zero Knowledge, with Kurt Hemecker @ Mina Foundation

- Building a Decentralized World of Knowledge Sharing and Data Verification, with Žiga Drev @ Trace Labs

- The Critical Importance of Building a Cloudless, Decentralized, Global Serverless Computing Network, with Tom Trowbridge @ Fluence Labs

- Leadership Insights from Building a Successful Full-Stack Blockchain Consulting Firm and Helping Clients Navigate the New Normal, with Vikram R Singh @ Antier Solutions

- How to Democratize Access to Money Issuance by Revolutionizing the Money Technology Stack, with Joao Reginatto @ M^ZERO Labs

- Why Multi-Signature Crypto Wallets Are Just a Band-Aid for a Much Larger Problem, with Simon McLoughlin @ Uphold

- $Y2 = X3 + 7$; Designing the Formula for the Future of Quantum-Resistant Computing, with Ian Smith @ Quantum EVM

- Crypto Hipster Presents...Shooting from the Hip! Ep 12: Why Designing Excellent Technological Standards Provides the Bedrock for Global Mass Adoption of Blockchain Technology; Fabian Vogelsteller @ LUKSO

- Quai, Qi, and the Serenity Effect of Transforming Compute Power into Money, with Alan Orwick @ Quai Network

- Building a Parallelized Confidential Computing Network that Helps Industries with Exposure to the Most Sensitive Data

Perform Encrypted Computations, with
Yannik Schrade @ Arcium

- Discovering Why Social Experiences are the
  Best Way to Onboard the Next Hundred
  Million People into Web Three, with Chris
  Liquin @ Cupcake

- Manifesting and Rewarding Attestations by
  Creating Customizable Social Graphs to
  Filter Your Proprietary Data, with Billy
  Luedtke @ Intuition

- Solving the Sovereign Trilemma and
  Improving Liquidity in Modular Ecosystems,
  w/ Karel Kubat @ Union

- Bringing Together Fragmented Web3
  Economies at the Southeast Asia Blockchain
  Week, with Nathan Kim @ UNOPND and
  Rachel Kim @ ShardLab

- Entering the Maitrix by Building an
  Economic Infrastructure for Decentralized
  AI, with Ian Estrada @ MAITRIX

- Why the Unsolvable Inefficiencies of Analog
  Traditional Financial Markets Open the Door
  for Digital Markets to Replace Them and
  Thrive in the Coming Years, with David
  Weisberger @ CoinRoutes

- A Hero's Journey: Reducing Fraud, Waste, and Theft and Transforming the Logistics Industry, with Todd Haselhorst @ HEALE Labs

- Why Smart Contracts and Data Capture Methods Must Continue to Evolve, with Nikhil Raghuveera @ Aethos

- The importance of capturing proof of authenticity while empowering a decentralized network of networks for the Web Three world, with Adam Helfgott @ Valence

Decentralized Finance (DeFi)

- The Future of Decentralized Finance is Here; It's Time for Regulators to Get on Board, with Hedi Navazan @ 1inch Group

- Why Consumer-Based DeFi Applications Are the Catalyst for Worldwide Blockchain Industry Adoption, with Cecilia Hsueh @ Morph

- Diving Into the "Down and Dirty" of Digging for Pristine DeFi Yield, with Nick Motz @ Soil

- How to Offer Decentralized Finance Yield-Earning Products in Emerging Market and High-Inflation Countries, with Max Galash @ Coinchange

- Redefining the Future of Decentralized Finance with Intents and Solvers, with Nikita Ovchinnik @ Barter DeFi

- Mixed Berries: Diving Into Decentralized Finance, Ethereum, NFTs, and More, with Jonathan Thomas @ Blueberry Protocol

- How to Build a DeFi Operating System that Drives Mainstream Adoption with Improved User Experiences, with James Lucas @ Lamina

- Discovering to Earn: Building Reward-Based Loyalty Programs; and the Importance of Recording Your Life's Achievements, with Jacob Margulies @ Galxe

- Building the Next Generation, Cross-Chain, Comprehensive Suite of Decentralized Finance Services, with Benedetto Biondi @ Folks Finance

- Pushing, Pulling, and Disentangling the Messy and Secretive World of Blockchain Oracles, with Chris Hermida @ Switchboard

- Crypto Hipster Presents: Reporter on the Ground, Episode 8; Exploring the Full Attestation of Blockchain Oracles, with Angus Tookey @ Chronicle

- Making DeFi Safer: How to Improve Cross-Chain Interoperability and Alleviate DeFi Vulnerabilities, with Mary McGilvray @ Interchain GmbH

- How to Develop a Single Ecosystem for Finance, Payments, and Trading Digital Assets, with Mike Romanenko @ Kyrrex

- How to Simplify Sophisticated DeFi Strategies for the Web3 World, with Kurapika @ Factor

- Exploring a Web3 OG Venture Capital Firm's Views on the Current Crypto Market and Decentralized Financial Landscape, with Alex Botte @ Hack VC

- Why Learning Chess Can Help You Rebalance Your Crypto Portfolio and Earn Greater Returns, with Danny Chong @ Tranchess

- Supercharging DeFi and Social AI Systems by Creating the Most Democratized Network, With Steven Pu @ Taraxa

- How Decentralized Networks Have Not Been Fully Decentralized and Why They Can Be Now, with Victor Vernissage @ Humanode.io

- Envisioning the Future of Decentralized Oracles and the Best Way to Help Traditional Financial Organizations Adopt Web3, with Marc Tillement @ Pyth Network

- Crypto Hipster Presents: Shooting from the Hip!, Episode 6: Why Smart Contracts Are Not Quite Smart Enough for the Mainstream DeFi User., with Adam Simmons @ RDX Works

- How to Help Startups and Institutions Empower Global Finance for the New Millenium, with Evan Owens @ Kadena

- How Restaking is Empowering an Open, Trustworthy Market and Why Crypto Regulators Should Take a 1990s Bill Clinton Internet Approach, with Warren Anderson @ Exocore

- How Syrup Can Achieve Stickiness in the Global Institutional DeFi Market, with Martin de Rijke @ Syrup, powered by Maple Finance

- How to Develop and Promote Blockchain Loyalty Reward Infrastructures, with Gabriele Giancola @ qiibee Foundation

- Designing Blockchain-Based Insurance Solutions to Mitigate Crypto Risks and Move Web3 Forward Powerfully, with Joseph Ziolkowski @ Relm Insurance

- Creating the First and Largest Liquid Staking and Re-staking Protocol in the Modular Ecosystem, with Josie Leung @ MilkyWay

- Building the First-of-Its-Kind, Permissionless DeFi Clearinghouse, with Barna Kiss @ Malda

- How the Web3 sector can combat the pertinent and costly issue of Maximal Extractable Value (MEV), with Da Hongfei, founder of Neo

- Overcoming the Challenges Facing Automated Market Makers and How to Avoid Unnecessary DeFi Costs, with Sunil Srivatsa @ Storm Labs

- Offering Sophisticated, Seamless, and Secure Innovations to Help Deliver Web3 Banking Solutions of the Future, with Myles Harrison @ AMINA Bank

Blockchain Education

- Why Raising the Benchmark in Blockchain Education is Important, with Cy Li @ Ethereum Collective Foundation

- Democratizing Blockchain Education as a First Step to Solving Web3 Developer Shortages, with David Bchiri @ XRPL Commons

Ethereum

- Leading the Charge to Safeguard and Enhance Ethereum's Historical Data Availability, with Ganesh Swami @ Covalent

- A Call to Action to Save Ethereum From Impending Disaster, with Alon Muroch @ SSV Labs

- Uncovering The Secret to Saving Ethereum Users Billions of Dollars in Transaction Fees, with Spring Dunn @ SKALE Labs

Fantasy Sports, Betting, and Entertainment

- Exploring the Future of Decentralized Sports Betting with Data-Driven Intelligence and Smart Leverage, with Mr. Blue @ LEVR.bet

- Why Karate Combat Is Appealing to the Younger Generations and How Fans and Influencers Can Drive the Future of Game Streaming Forward, with Kyle "Post Master" @ Karate Combat

- Re-imagining Sports Book Betting by Enabling Users to Be the House, with Peter Argerakis @ Six Sigma Sports

- Helping Brands Take Control of Their Narratives Through Fandom-Driven Experiences, with Anthony Rodriguez @ The Digital Spenders Club

- Creating the First Web3 Fantasy League for Movie Fans, with Stacy Spikes @ Mogul by MoviePass

- Creating an Online Casino at the Crossroads of Crypto and Culture, with Zach Bruch @ MyPrize
- Shark, Pickle, Cone Documentary Movie to Debut on May 29th at Consensus 2024, with

Chris Waters and Neil Berkeley @ Stoopid Buddy Stoodios

Web3 Gaming

- Challenging the Public Perception of What Web3 Gaming Should Be, with Max Fu @ Nyan Heroes

- The Challenges and Experiences Behind Building the World's Largest Social and Casual Web3 Game, with Luke Barwikowski @ Pixels

- Crypto Hipster Presents: Reporter on the Ground, Episode 2; How to Help Gaming Studios Stop Flying Blind, with Lucas Fulks @ Helika

- Still Focusing on the Fun: Helping Developers, Content Creators, and Game Players Build the Future of Web Three Gaming, with Marc Mercuri @ Shrapnel

- How to Transform the Gaming and Advertising Industries With Decentralized Payment Solutions, with Timothy Tello @ 3thix

- Creating a World Development Gaming Studio at the Intersection of Blockchain and AI, with Ilman Shazhaev @ Dizzaract
- Creating a Fun Ultimate Playground for Gamers to Unlock Web3 Experiences, with Daniel Anthony @ ZKcandy

- Transforming Normies into Degens with Progressive Web App Games, with Tomer Pascal @ OwnPlay

- Slaying the Bear and Building the Future of Sustainable Web3 Strategy Gaming, with David Johansson @ BLOCKLORDS

Identity, Storage, and Commerce

- How to Drive Innovative Breakthroughs in the Decentralized Digital Identity Space, with Harrison Seletsky @ SPACE ID

- Examining the Current Decentralized Storage Landscape to Help AI Fulfill Its Potential in Web3, with Ryan Levy @ DataHaven / Moonbeam

- Discovering the Parallels Between Blockchain Advancements and the Evolution of Multi-Channel E-Commerce, with Neil Twa @ Voltage Holdings

- Combatting AI-Driven Fraud Using Crypto File Storage Solutions, with Kyle Tut @ Pinata

Intellectual Property / Thought Leadership

- How to Discover Market-Redefining Opportunities and Support Emerging Entrepreneurs in Web3 and Beyond, with Jameel Qeblawi @ myqubator

- All Aboard the Ideas Sled!: Leveraging Innovative Technology to Inspire and Influence Change with Grace, with Christopher G. Fox, PhD @ Ideas-Led Growth

- Crypto Hipster Presents: Reporter on the Ground, Episode 6; Discovering How a Cookie Can Disrupt the Global Digital Marketing Industry without Crumbling, with Krystyna Kozak-Kornacka @ Cookie3

- How to Create Your Self-Sovereign Social Identity with Proof of Humanity, with Lasha Antadze @ Rarify Labs (Part 2 of 2)

- How to Build a Brand People Can Trust in Nascent Industries, with Armel Leslie @ RF|Binder

- How to Position Your Web3 Brand at the Forefront of the Real-World Asset (RWA) Movement, with Bhaji Illuminati @ Centrifuge

- Telling Stories, Building Trust, and Helping Web3 Brands Succeed, with Samantha Yap @ YAP Global

- Discovering a World-Building, Immersive Platform for Creators by Creators, with Justin Melillo @ MONA

- Designing a Revolutionary Platform to Transform How Intellectual Property Ownership Is Created, Shared, and Monetized on Web3, with Jaime Schwarz @ MRKD

- Helping Web3 Micro-Influencers and Small Creative Agencies Gain Massive Audiences, with Ryan Davis @ People First
- Realizing the Benefits of Treating Your Web3 Community as CEO, with Kelsey McGuire @ Shardeum

- Building an Avant-Garde Social Gaming Platform that Empowers the New Creator Economy, with Casey Grooms @ Soulbound

Litecoin

- Embracing a Brighter Vision for the Future of Litecoin and Its Global Adoption in 2025 and Beyond, with David Schwartz @ Litecoin Foundation

Macroeconomic Outlook

- Why the Macroeconomic Outlook for Crypto in the U.S. is Extremely Bullish Despite Recent Short-Term Market Volatility, with Kyle Reidhead @ Milk Road

- Crypto Hipster Presents...Shooting from the Hip! Episode TEN. Insights from Argentina: Marrying Traditional Finance's Stability with Blockchain's Innovation Potential, with Agustin Liserra @ Num

- Crypto Hipster Presents...Shooting from the Hip! (E15): Why 2024 Will Be the Year that Crypto-Focused Voters Tip the U.S. Election Results in Pro-Blockchain Candidates' Favor, with Matthew Le Merle

- Crypto Payment Insights and Predictions for the Upcoming Bull Market, with Sean Mackay @ CoinPayments

- How SHOPX is Empowering Customer Ownership in Web 3, and How to Navigate this Current Inflationary Period Successfully, with TJ Chang @ SHOPX.

Mobility

- How to Revolutionize the Ride-Sharing Economy Through Web3 Innovation, with Firdosh Sheikh @ DRIFE

- Crypto Hipster Presents...Shooting from the Hip! (Ep 9): Shaping the Decentralized Future of Mobility Markets, with Deniz and Marko @ Soarchain

Non-Fungible Tokens (NFTs)

- Why NFTs May Still Just Be the Catalyst That Brings Billions of Web2 Users into Web3, with Aka Leung @ Bitget

- Season 7 Premiere: Solving Experiential and Developer Challenges with Web Three Gaming, Digital Collectibles, and NFTs, with Dr. Alun Evans @ LAOS Network

- Solving the Greatest Challenges Facing the Cannabis Industry with NFTs, with Ricardo Capone @ Dr. Green NFT

- Breaking Down Ivory Towers and Stone Walls: How NFTs Can Enhance Life Experiences and Benefit Local Communities, with Tyler Adams @ COZ

- Alive and Kickin': Why NFTs Are Far From Dead; Rather, an NFT-Future Looks Quite Bright, with Rusty Matveev @ Calaxy

- Revolutionizing the Best of Both Traditional Art and Digital Art Worlds with Digital Twin NFTs [SEASON 6 FINALE], with Zain Talyarkhan @ Notable.art

Regulatory and Compliance

- How to Blend Fiat and Crypto Payments in a Secure and Compliant Way, with Nikolay Denisenko @ Brighty

- How to Navigate the Global Crypto Regulatory Environment Successfully, with Norman Wooding @ SCRYPT

- Cutting through the Bullsh*t U.S. Regulatory Hurdles to Help Provide Your Institution's CFO Office with Crypto Financial Reporting Solutions that Genuinely Work, w/ Amy Kalnoki and Pat White @ Bitwave

- Why KYC, AML, and Black Swan Insurance are Essential Pillars to Onboard Institutions into Decentralized Protocols, with Ramon Recuero @ Kinto

- Helping Virtual Asset Service Providers Navigate the New Travel Rule Requirements Globally, with Emeka Mgbenu @ Sumsub

- Why Using the Telegram Platform for Compliant Digital Asset Trading is the New Frontier, with Chilip Lai @ LeapXpert

- Helping Enterprises Navigate MiCA by Providing a Comprehensive Solution for

Managing Digital Assets, with Tom Kiddle @ Palisade

- Tackling the Complexities of Legally Compliant Fungibility for Stablecoin Issuers in Europe Under MiCA Regulations, with Gijs Op De Weegh @ StablR

Real-World Asset (RWA) Tokenization

- Overcoming the Challenges of Tokenizing and Fractionalizing Real-World Assets, with Graeme Moore @ Polymesh Association

- Why the Tokenization of Real-World Assets Will Front-Run Traditional Finance, with Niklas Kunkel @ Chronicle Labs

- How to Bring Real-World Tokenized Assets (RWAs) to the Masses, with Chris Yin @ Plume

- Tokenizing the World: Why the Future of Crypto and Blockchain Technology is the Mass Tokenization of Real-World Assets, with Carlos Balbin

- Designing the First Genuinely Global Tokenized Infrastructure and Decentralized Real World Asset (RWA) Marketplace, with Miguel Buffara @ RACE

- The Importance of Creating a Multi-Chain Asset Tokenization Launchpad that Tokenizes Any Asset, with Jeroen Offerijns @ Centrifuge

- Crypto Hipster Presents: Shooting from the Hip!, Episode 4: The Impending Rapture for Tokenizing Real-World Assets and the Resurrection of Satoshi Nakamoto's Original Bitcoin Vision, with Jonny Fry

- Real World Assets, Democratizing Access to Data, and the Advantages of Decentralization, with Stefan Rust @ Truflation

- Bringing Real-World Assets (RWAs) On-Chain for Low-Cost Financing and Regenerating the NFT Market, Kkrusher (Kevin Rusher) @ RAAC

- Unlocking Tokenized Ownership in the World's Most Luxurious Resorts, with Ricardo Johnson @ Oases

- How to Leverage Blockchain Technology to Secure Smooth, Efficient, and Trustworthy Real Estate Transactions, with Erik LaPaglia @ Propy

- Using the Internet of Things, Machine Learning and Blockchain Technology to Help Farmers in Marginalized Areas, with Jon Trask @ Dimitra

- Tokenizing GPUs and Their Yields to Create a New Asset Class for AI and Compute, with Kony @ GAIB

- How to Transform Access to Compute by Creating GPU Financial Assets, with Nikolay Filichkin @ Compute Labs

- How to Transform Idle Tokens Into Productive Assets, with Altan Tutar @ MoreMarkets

- Transforming Real-World Compute Into On-Chain Assets Through Mining Hardware Tokenization, with Leo Fan Xiong @ Cysic

Social Impact

- Discovering a New Currency, a New Financial Freedom, and a New World, with Dr. Karl Huber @ Quai Network

- Decentralization: The Bridge to Abundance for Humanity, with Violet Abtahi @ Platonic

- Girl Talk, Taking Risks, and Creating Timeless Classics in a Blockchain World, with Claire Ross-Brown @ CJ London

- "I Have a Dream!": Building a Valuable Global Economy and Collaborative World Society Together, with Web3 Technology, Ethos, and Values, with Anjali Young @ Collab.Land

- Reforming Socio-Economic Systems to Facilitate the Generation and Spread of Natural Wealth via Energy, Finance, and Bitcoin Mining, with Mohamed El-Masri @ Hodler

- Exploring Humanity's Role in a Super-Intelligence AI-Driven Future, with Jamie Goldblatt @ Mind Chill AI—Cyberverse of Chill

- 4th of July 2024 Special: How to Chart Your Personal Independence Into the Web Three World through AI and Blockchain Technology, with J.D. Seraphine @ Raiinmaker

- Recognizing and Celebrating Innovative Technologies That Can Improve Our Daily Lives and Ensure America Will Thrive in the Modern Digital Arena, with Joy Schoffler @ Distinctive Edge Partners

- Crypto Hipster Presents: Shooting from the Hip, Episode 2: Promoting Sustainable Development and Economic Empowerment in Ghana, with Ibrahim Mustapha @ Me for Africa

- Crypto Hipster Presents: Shooting from the Hip! (E5): How a Tech Maverick with an Unwavering Commitment to Preserving People's Rights to Privacy is Reshaping the Global Digital World, with JB Benjamin

- Solving the Current Challenges in Using Crypto as a Payment Method for the World's Spanish-Speaking Population, with Javier Castro-Acuña @ Bitnovo

- Empowering Users and Marginalized Communities by Sharing Hotspot

Connections and Enabling Access to Web Three, with Ugochukwu Aronu @ Wicrypt

- Helping Retail Consumers Thrive by Transforming Web3 Ecosystems to Adopting an Inclusive Economic Model Instead of a Player-versus-Player Trap, with Cyrus Taghehchian @ Aces.Fun

- Crypto Hipster Presents: Reporter on the Ground, Episode 7; Creating Collaborative Innovation through Social Capitalism, with John Wingate @ Bank Social

- Crypto Hipster Presents: Reporter on the Ground, Episode 9; Creating a New Unified Experiential Society, with John Vibes @ Somnia

- Charting New Paths for Humanity to Preserve the Potential of People Amidst an Ever-Increasing Artificial Intelligence Revolution, with Terence Kwok @ Humanity Protocol

Solana

- How to Design the Most Flexible and Efficient Staking Strategies and Experiences for Solana Users, with Michael Repetny @ Marinade

- Insights From Building Solana's First Meta Decentralized Exchange Aggregator, with Chris Chung @ Titan

- Why Ethereum Must Be Saved Soon, and How the Solana Virtual Machine Can Help, with Joanna Zeng @ SOON

- Advancements in Oracles and the Evolution of Solana Beyond Simply "Sam's Coin", with Mitchell Gildenberg @ Switchboard

Sustainability

- Leveraging Distributed Ledger Technology to Create ESG Impact Initiatives that Facilitate Sustainability, Reliability and Integrity, with Andrew Forson @ The Hashgraph Association

- Solving Greenwashing, Environmental Propaganda, and the United Nations' Biodiversity Challenges with Generational Blockchain Technology, with Owen O'Driscoll @ Trrue

- How to Build a Decentralized Movement for Emissions Reporting to Ensure Accurate Business Sustainability Claims, with Caitlin Moore @ Filecoin Green

- How the Intersection of AI, ESG Investing, and Carbon Credit Tokenization Can Help the United Nations Meet Their Sustainability Goals, with Daniel Steeves

The Digital Economist

- How to Leverage Blockchain and Sustainable Innovative Technology to Drive Economic, Environmental, and Social Advancements Globally, with Dr. Nikhil Varma

- How Socioeconomic Forces Driving an AI Revolution Will Lead to Advancements in Human Evolution, with Dr. Shruti Shankar Gaur @ The Digital Economist

- Guest Episode 500: Innovation, Inspiration, and the Rise of the Tech-Infused Entrepreneur, with Jean Criss @ The Digital Economist; Jean Criss Media; CRISSCROSS Intimates

- How to Nationalize Digital Currencies, With Lessons From Implementing Chivo Bitcoin Wallet in El Salvador, with Tristan Thoma @ Impera Strategy

Appendix D: Crypto Hipster's Mysticals Books on Amazon.com

These books are the Crypto Hipster Mysticals episodes from Seasons Four and Five, as published in book format on Amazon. They are compilations of three or more episodes captured primarily for readers and historians. There are twenty-two books in the Crypto Hipster's Mysticals compilation, with another seventy-three plus books under various collections.

- The Sword of Ja-million: The Future of Web Three Gaming in a Globally Decentralized World (Crypto Hipster's Mysticals Book 1)
- Savory Wild RICE: Rights, Inspiration, Currency, and Enthusiasm for the People with Blockchain Technology (Crypto Hipster's Mysticals Book 2)
- Unreal, Surreal, or Artificially Real?: Influencing the World Through a Rapidly Growing Artificial Intelligence and

Blockchain Technology Landscape (Crypto Hipster's Mysticals Book 3)

- Sugary Maple Eggs and Scrambled Bacon: The Future of Crypto Platforms, Infrastructure, and Decentralized Autonomous Organizations (Crypto Hipster's Mysticals Book 4)
- Hot Steamy Luggage: How Artificial Intelligence, Blockchain, and the Metaverse Can Restore Humanity (Crypto Hipster's Mysticals Book 5)
- Hot Spicy Luggage: Regulatory Affairs, Attacks on Privacy, and Why Governments Need Central Bank Digital Currencies (Crypto Hipster's Mysticals Book 6)
- Hot Smoky Luggage: A Spot Check Inventory on Crypto Market Prices From 2023 (Crypto Hipster's Mysticals Book 7)
- Hot Sapid Luggage: Ushering in a Prosperous Decentralized Finance Future (Crypto Hipster's Mysticals Book 8)
- Hot Scrumptious Luggage: Sustainability and Dominion: Bridging the Divide Between the Abstract and the Concrete (Crypto Hipster's Mysticals Book 9)
- Not Just All About Cats, Dogs, Sloths, and Other Fuzzy Digital Creatures, Volume One: The Staggering Distance Between a JPEG and a Carbon Copy of Reality (Crypto Hipster's Mysticals Book 10)
- Not Just All About Cats, Dogs, Sloths, and Other Fuzzy Digital Creatures, Volume Two:

Opportunities for Self-Sovereign Web3 Journeys (Crypto Hipster's Mysticals Book 18)

- Let's Play Checkers; Let's Play Risk...Yes, Yeh, Yea, Yeah!!: Overcoming Challenges Facing Communities in Decentralized Social Finance, Gaming, and Web3 Media (Crypto Hipster's Mysticals Book 19)
- Choosing Peace in a Turbulent World: How Cryptocurrencies Can Serve As a Safe Haven During Challenging Political Times (Crypto Hipster's Mysticals Book 20)
- A Father's Cry For Help: How a Courageous Father Persevered Against an International Criminal Conspiracy to Seek Justice for His Son (Crypto Hipster's Mysticals Book 21)
- Everyday People: The Web3 Version: Our Global, Social, Financial, and Economic Future is Decentralized (Crypto Hipster's Mysticals Book 22)

Appendix E: Crypto Hipster's Chronicles Books on Amazon.com

These books are the Crypto Hipster Chronicles episodes from Seasons One, Two, and Three, as published in book format on Amazon. They are compilations of three or more episodes captured primarily for readers and historians. There are forty-six books in the Crypto Hipster's Chronicles compilation, with another two-hundred plus books under various collections.

- Old Paradigms Die Hard: Proof of Resilience: How Ukraine Built a Sovereign Digital Economy in the Face of War with Russia (Crypto Hipster's Chronicles Book 1)

- From Stuck to Unstoppable: How Blockchain Technology Can Help You Overcome Adversity, Depression, and Suicidal

Thoughts (Crypto Hipster's Chronicles Book 2)

- Why Wall Street Is Running Scared: How Decentralized Finance Has the Titans of Wall Street Shaking in Their Suits (Crypto Hipster's Chronicles Book 3)

- Origins of the Metaverse: How the Metaverse is Metamorphosizing the Future of Digital Technology (Crypto Hipster's Chronicles Book 4)

- An Altruistic New Paradigm: The Pursuit of Trust and Altruistic Good (Crypto Hipster's Chronicles Book 5)

- A State in a Smartphone: A Safe Harbor for Ukraine (Crypto Hipster's Chronicles Book 6)

- Play Amazing Games, Earn Amazing Prizes: The Play-to-Earn Benefits of Crypto Gaming (with Ancients and Heroes) (Crypto Hipster's Chronicles Book 7)

- The Zoology and Cheese of NFTs: The Many Shapes and Sizes of Non-Fungible Tokens (NFTs)... For You's and For Me's (Crypto Hipster's Chronicles Book 8)

- The Crypto Nomad's Powerful Brew: How Decentralization Combats Frauds, Rugs, and Digital Crime (Crypto Hipster's Chronicles Book 9)

- Sold Out in Fifteen Seconds: NFTs for Gaming, Art, Investments, Graffiti, and Digital Disneyland (Crypto Hipster's Chronicles Book 10)

- Knowledge is Power: How The Knowledge Society Helps Students Make a Significant Impact in the World (Crypto Hipster's Chronicles Book 11)

- Trees, Birds, and Fire: How Musicians and Artists Can Leverage Web 3.0 to Build Fan Engagement (Crypto Hipster's Chronicles Book 12)

- Cats and Cancer, Iron and Power: How NFT Artists and Entrepreneurs are Overcoming Adversity and Building Resiliency with Blockchain Technology (Crypto Hipster's Chronicles Book 13)

- The Crypto Nomad's Galactic Journey: Scooters, EVs, and Outer Space with Blockchain, Sustainability, and Hustle, Baby!! (Crypto Hipster's Chronicles Book 14)

- Metaverse Challenges Beneath Z Hype: Exploring Research, Development, and Logistics of Crypto Gaming and the Metaverse (Crypto Hipster's Chronicles Book 15)

- My Web3 NFT Went to the Market: Early Web3 and NFT Marketplaces and Healthcare

Benefits While Not Staying Home (Crypto Hipster's Chronicles Book 16)

- Fierce... Courageous... Heroes: Creating a World of Heroes with Bitcoin and NFTs (Crypto Hipster's Chronicles Book 17)

- Mitigating Crypto Control Risks: Addressing Crypto Compliance and Global Regulatory Standards (Crypto Hipster's Chronicles Book 18)

- The Digital Path For Global Sustainability Goals: Achieving Global Sustainability Goals through Bitcoin Mining and Community Coins (Crypto Hipster's Chronicles Book 19)

- Valley and Thunder; Springtime and Crypto Winter: Early Predictions and Trends from 2022 (Crypto Hipster's Chronicles Book 20)

- Privacy, Data Ownership, and Freedom in the Web Three Era: Controlling and Securing Your Personal Web 3 Data and Online Identity (Crypto Hipster's Chronicles Book 21)

- Breaking: Wall Street; Why: Bitcoin: Why Bitcoin is Impervious to Overreaching Bureaucrats and Unethical Politicians (Crypto Hipster's Chronicles Book 22)

- Cash Flows in Crypto Valley: How Switzerland is Poised to Become the Global

Crypto Leader (Crypto Hipster's Chronicles
Book 23)

- Governance, Intelligence, and Justice: Three
  Pillars of Decentralized Autonomous
  Organizations (Crypto Hipster's Chronicles
  Book 24)

- Solving the Carbon Neutrality Crisis:
  Discovering Bitcoin Mining's Beneficial
  Impact for the Environment (Crypto
  Hipster's Chronicles Book 25)

- Bitcoin Ethics in a Corrupted World: Are We
  Achieving Satoshi Nakamoto's Vision or
  Simply Succumbing to Unethical
  "Regulators"? (Crypto Hipster's Chronicles
  Book 26)

- Persevering Greatly in an Ocean of
  Unqualified Critics: Building a Resiliency
  Skill Set through Recreating Memories,
  Empowering Women, and Fighting for ...
  World (Crypto Hipster's Chronicles Book 27)

- War... and Peace... and... Altruism: War
  Games, Utility Tokens, AK47s, and
  Blockchain for Peace (Crypto Hipster's
  Chronicles Book 28)

- The Futurrr(e-2r) of Finance: DeFi-N-ing the
  Future of Finance for Widespread Crypto
  Adoption (Crypto Hipster's Chronicles Book
  29)

- Stompin'!... by a Tree Trunk on a Snowy Evening: Uplifting Creators and NFT Collectors through Cutting-Edge Technologies and Zero-Knowledge Innovation (Crypto Hipster's Chronicles Book 30)

- Crypto: The Empire of the Shun?: Ethics, Conviction, and Resiliency—Important Personal Traits or Blockchain Cornerstones? (Crypto Hipster's Chronicles Book 31)

- Oogie! Groovy!! Boogie!!! DeFi Inferno!!!!: Funds, Crypto Banks, and Alternative Investments: Doing Decentralized Finance (DeFi) in Different Ways (Crypto Hipster's Chronicles Book 32)

- Burn Rubber and Chill Your Mind: Creating Equitable Futures and Healthy Ecosystems with Crypto Tokens, NFTs, and Metaverses (Crypto Hipster's Chronicles Book 33)

- Human Nature: Hipsters, Heroes, and Handbooks: Discover Your Inner NFT in the Real World and in the Metaverse (Crypto Hipster's Chronicles Book 34)

- Crypto Hipster: Crypto Hipster's 100th Episode: Untold Insights with Bitcoin Legend Charlie Shrem (Crypto Hipster's Chronicles Book 35)

- Sweeter Than a Reggie Bar: A Critical Look for Us Beyond Just Web Three (Crypto Hipster's Chronicles Book 36)

- The Superfreaky Groove Is in Your Heart: Starting from Zero and Building a Powerful Personal Blueprint for Success (Crypto Hipster's Chronicles Book 37)

- Boys to Chads: Motown Cryptodelphia: How the Blockchain Helps Content Creators Build Bright and Abundant Futures (Crypto Hipster's Chronicles Book 38)

- 100X Marks the "Oh, Gee!" Spot: Pitfalls and Rewards of Crypto Trading and Investing (Crypto Hipster's Chronicles Book 39)

- Sashay, Shante, & Prêt-À-Porter: (Ransomware, Gem Wear, and Underwear): Measuring Blockchain's Beneficial Impact on Disparate Industries (Crypto Hipster's Chronicles Book 40)

- The Rejuvenating Power of Sweet, Sweet Doctor Love: Empowering Healthcare Participants and Real Estate Owners in the Web 3 Economy (Crypto Hipster's Chronicles Book 41)

- Purply Fire, Orangey Rain, and the Funky Green Sandstorm of Freedom: Crypto Across Four Corners: Blockchain Insights from Around the World (Crypto Hipster's Chronicles Book 42)

- Securities, Tunnels, and the Surprisingly Delightful Splendor of Entropy: Creating a Path Forward in Decentralized Finance for Global Banks and Institutions (Crypto Hipster's Chronicles Book 43)

- The Unequivocal Usurpingly Manifesting Powerful Strength of the Spec..., Spec..., Spectacularly Spectacular Triangle Mindset (Crypto Hipster's Chronicles Book 44)

- Golden Fire in Your New Bitcoin Shoes: Why Economists and Environmentalists Are Often Wrong About Bitcoin (Crypto Hipster's Chronicles Book 45)

- Block Berry, Block Banana, and Block Sky Action: The Digital Crypto Future is Coming Sooner Than You Think (Crypto Hipster's Chronicles Book 46)

Appendix F: Full Season Crypto Hipster Compilations and Other Notable Books by the Author

Crypto Hipster's Full Season Compilations

Crypto Hipster: Crypto Hipster Podcast: The Complete First Season (March 1, 2021—June 30, 2021) (Crypto Hipster's Full Season Compilations)

Crypto Hipster: Crypto Hipster Podcast: The Complete Second Season (August 1, 2021—December 31, 2021) (Crypto Hipster's Full Season Compilations)

Crypto Hipster: Crypto Hipster Podcast: The Complete Third Season (January 1, 2022—June 30, 2022) (Crypto Hipster's Full Season Compilations)

Crypto Hipster: Crypto Hipster Podcast: The Complete Fourth Season (August 31, 2022—December 31, 2022)

Crypto Hipster's Golden Splendor: Crypto Hipster Podcast: The Complete Fifth Season (January 1, 2023—August 19, 2023)

Blockchain Ethics

Blockchain Ethics: A Bridge to Abundance

Blockchain Ethics: Arise from the Ashes

Blockchain Ethics: Fighting Honorable Battles

Generation X Manifesto

Re-Generation X: How Generation X Can Leverage Blockchain Technology to Save Themselves and Rebuild America

Appendix G: Author Jamil Hasan in the Media

Jamil Hasan - Crypto Planet (11/8/2017)

https://www.youtube.com/watch?v=eWiI52yTyes

Jamil Hasan Live from Studio 6B (3/28/2018)

https://www.youtube.com/watch?v=y3cTVoedFsc

Jamil Hasan Daily Magic Interview (7/12/2018)

https://www.youtube.com/watch?v=1fHrT8dIg5g

Jamil Hasan Spotlight TV Interview (7/28/2018)

https://www.youtube.com/watch?v=OubIQ5G6gq
w

Naked Marketing #028 - Jamil Hasan - Marketing Mistake - Switching strategies too often (9/22/2021)

https://www.youtube.com/watch?v=_U_T3D5Izc
M

**METAVERSE INTERVIEWS with LeDrop
WithCheese - Episode 4 w/guest Jamil
Hasan, blockchain author and thought
leader** (7/28/2022)

https://www.youtube.com/watch?app=desktop&v=
qwss7YxXzug

**The Crypto Corner podcast with Jamil
Hasan: Moving Hemp Forward** (8/17/2022)

https://www.youtube.com/watch?v=FEgmftGQ19I

**Conversation with the Publisher of Crypto
Hipster Publications, Jamil Hasan, on the
Influencers Podcast with Dr. Bill Williams**
(1/23/2023)

https://www.youtube.com/watch?v=3D8Z7IUovco

**S2E2 Crypto With English with Jamil Hasan,
DeFi Author & Host of the "Crypto Hipster"**
(6/9/2023)

https://www.youtube.com/watch?app=desktop&v=
Nfsokpkc_e4

ThriveMore with Roger Martin: E138: Jamil Hasan: Crypto Investing 101: Making Sense of Blockchain and Bitcoin (7/2/2025)

https://www.youtube.com/watch?v=V2Q5V8PSSxQ

Made in the USA
Monee, IL
07 July 2026

56544292R00121